Practical Exercises in Microelectronics

Also from Stanley Thornes

MICROELECTRONIC SYSTEMS – G.M. Cornell
MICROPROCESSOR INTERFACING – G. Dixey
MICROELECTRONICS NII – D. Turner
MICROELECTRONICS NIII – D. Turner

Practical Exercises in Microelectronics

David Turner BSc(Eng) CEng MIEE ACGI

Dean of the Faculty of Technology
College of Further Education, Plymouth

Stanley Thornes (Publishers) Ltd

First published in 1992 by:
Stanley Thornes (Publishers) Ltd
Old Station Drive
Leckhampton
CHELTENHAM GL53 0DN
England

British Library Cataloguing in Publication Data
Turner, David
    Practical exercises in microelectronics.
    – (Microelectronics series)
    I. Title II. Series
    621.381

    ISBN 0–7487–1305–0

Typeset by Florencetype Ltd, Kewstoke, Avon
Printed and bound in Great Britain at The Bath Press, Avon.

Contents

Foreword

In 1982 Hutchinson Educational Publishers, now part of Stanley Thornes, published on behalf of the Business Technician Education Council (BTEC), a series of books designed for use as learning packages in association with the published standard units in Microelectronic Systems and Microprocessor-based Systems.

The last decade has seen the transformation of industrial computing with the explosion in personal computing. The reduction in price complemented by a significant increase in computing power has extended the application of personal computers so that microelectronics is now a realistic tool in all sectors of industry and commerce. The need for adequate training programmes for technicians and engineers has increased and the BTEC units have been revised and updated to reflect today's needs.

Stanley Thornes have produced a series of learning packages to support the updated syllabuses and numerous other courses which include microelectronics and Microprocessor-based Systems. There are five books in the series:

Microelectronic Systems	Level F	by G. Cornell
Microelectronics NII	Level N	by D. Turner
Microelectronics NIII	Level N	by D. Turner
Microprocessor-Based Systems	Level H	by R. Seals
Microcomputer Systems	Level H	by R. Seals

Two additional books which complement the five above are:

Microprocessor Interfacing	Level N	by G. Dixey
Practical Exercises in Microelectronics	Level N	by D. Turner

This book is a series of practical exercises which complement the two Microelectronics Level NII and Level NIII books with a programme of relevant laboratory-based exercises. The three books, therefore, give a complete theoretical and practical coverage of N level microelectronics and microprocessor systems.

Andy Thomas
Series Editor

Preface

This book is written to cover the practical objectives of the BTEC Unit Microelectronic Systems U86/333 at both N2 and N3 level. The companion volumes in the series cover the necessary microelectronics theory which should be studied at the same time as this practical work. It provides a large number of relevant practical exercises designed specifically for student use which become progressively more complex throughout the book. The three Microelectronics books cover all aspects of the BTEC Unit from a theoretical and practical point of view, providing ideal source material for both classroom activities and self-supported study.

For those students who prefer to study by distance or open learning methods, two learning packages are available which integrate the theory and the appropriate practical activities. The packages also contain all the required hardware and software needed for a complete microelectronics course. Details are available from:

Plymouth Open Learning Systems Unit,
College of Further Education,
Kings Road,
Devonport,
Plymouth PL1 5QG
Telephone: (0752) 551947
Fax: (0752) 385343

Any practical book must be based on specific equipment to be of any immediate value. In this case, the practical exercises are written for the Multitech Micro-Professor MPF-1B, together with its applications board and single step board.

There are a number of reasons for this.

(a) The Micro-Professor is already very widely available throughout the education and training institutions of the United Kingdom. It is unlikely that there are many establishments who have never used this particular single-board computer.

(b) It is an adaptable system so that it is very simple to add the peripheral boards such as the applications board and the single step board. This makes it ideal for introductory microelectronics practical exercises.

(c) It is Z80 based and this means that it can be used to complement the theory elements of the other microelectronics books in the series.

(d) The system is relatively cheap, so that anyone contemplating starting to learn microelectronics from scratch can do so relatively cheaply.

A full introduction to the Micro-Professor and the applications board is given in the first chapter of this book. To obtain maximum benefit from the practical exercises each student should have access to:

(a) Multitech Micro Professor MPF-1B.
(b) Micro-Professor applications board.
(c) Micro-Professor single step board.
(d) Z80 Cross-assembler (for BBC or IBM PC) and connecting lead.
(e) POLSU replacement EPROM set.
(f) Practical exercises on disk (IBM PC or compatibles only).

Items (a)–(d) are available from:

Flight Electronics Ltd.,
Flight House,
Ascupart Street,
Southampton SO1 1LU
Telephone: (0703) 227721
Telex: 477389 FLIGHT G
Fax: (0703) 330039

Items (a)–(f) are available from Plymouth Open Learning Systems Unit (POLSU) whose address was given earlier.

Most of the exercises can be performed equally successfully using the Micro-Professor MPF-1P, together with the applications board and single step board.

The aims of this book are to introduce the student to the practical aspects of microcomputer hardware, software and applications. It includes about 30 exercises which may be performed by students to support the theoretical aspects of micro-computer systems. They cover topics such as system architecture, tracing program operation, mathematical applications, control systems, sequencing, peripheral operation, analogue and digital input/output and the use of interrupts. All exercises are well tried and tested and come with complete software listings.

David Turner
October 1991

Acknowledgements

The author wishes to thank his wife Kathy, whose help and encouragement during the production of this book have been a constant source of inspiration and whose sense of humour has somehow managed to remain intact throughout. Thanks also go to his understanding children Beth, Alex and Ben, who should have seen a lot more of their father than they have done.

His thanks go to Jackie Boyce who typed and edited the manuscript so efficiently and whose typing speed is a good test for any wordprocessor. Thanks also go to Roger Bond and Elizabeth Frederick-Preece of POLSU who created some excellent illustrations from his very rough sketches.

He is grateful to Max Soffe of Flight Electronics Ltd. for permission to include the Micro-Professor circuit diagrams, and also to Acer TWP Inc. whose design it is.

He is also grateful for the help of Trevor Burrows (Professional Photography) in the production of the photographs used in the book.

ONE

Introduction to the computer system

EQUIPMENT REQUIRED

To complete all of the practical exercises in this chapter, you will need:

(a) *Micro-Professor MPF-1B.*
(b) *Applications board MAB.*
(c) *The POLSU replacement EPROM set (installed).*

PREREQUISITES

Before studying this chapter you should have:

(a) *An understanding of the basic components of a microprocessor based system.*
(b) *An appreciation of the hexadecimal number system.*
(c) *An awareness of the different types of memory and their address ranges.*

These may be obtained by studying the first chapter of the companion book in the series, *Microelectronics NII.*

1.1 THE EQUIPMENT

Most of the practical exercises in this book require only three items of equipment:

- The Micro-Professor MPF-1B.
- The applications board MAB.
- The POLSU replacement EPROM set.

The applications board is attached to the PIO of the Micro-Professor with a 40-way ribbon cable, and the POLSU EPROM set is used to replace these EPROMs which are supplied with the Micro-Professor.

In addition, some exercises use the Micro-Professor single step board which allows the computer to step through programs one machine cycle at a time.

Figure 1.1 (overleaf) shows this complete system.

All of the exercises have been written so that they could be entered directly in hexadecimal code on the Micro-Professor keyboard. In some cases this is the simplest and quickest method of program entry and details of how to do this are supplied with the Micro-Professor.

However, some students may have access to an IBM-compatible PC, and by using a Z80 cross-assembler it is possible to use the PC to write the programs then simply download them into the Micro-Professor. The POLSU Replacement EPROM set contains the required software to receive such programs. This allows the programming to be done in assembly language which is much easier, and also allows the software to be retained on disk.

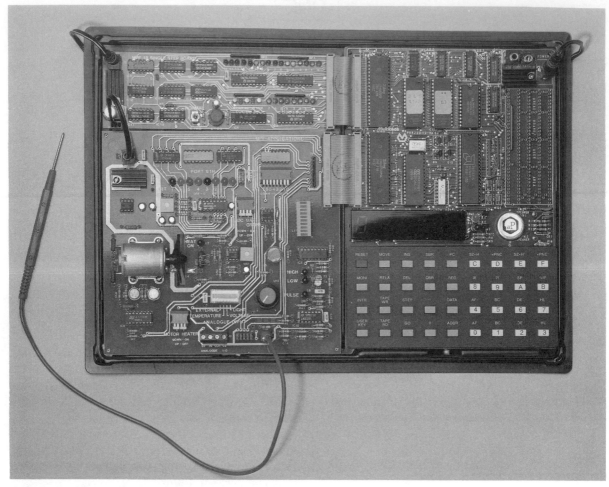

Figure 1.1   Microcomputer system

Students who wish to use assembly language should refer to the practical exercise in the Appendix which gives details of how the cross-assembly and downloading process takes place. The programs are also available ready assembled and as source code on disk for those who do not have time to type them all into the computer.

1.2 FEATURES OF THE MICRO-PROFESSOR

The Micro-Professor MPF-1B is a Z80 based single board computer which operates at a clock speed of 1.79 MHz. It has space for 8K of ROM in two sockets which occupy addresses 0000–0FFF hex and 2000–2FFF hex, respectively.

It can have up to 2k bytes of RAM which is located at addresses 1800–1FFF hex.

The display is a six-character, seven-segment LED array driven from an Intel 8255 parallel input/output chip. This is also used for the 36-key keyboard interface. All of the control for both keyboard and display is achieved in software.

On board there is also an uncommitted Z80 PIO chip and a Z80 CTC. These provide ideal interfaces for the applications board which is used in most of the following practical exercises.

A tape interface is also provided in the main board which can be used with a cassette recorder. However, the input may also be used to receive data from another computer, and this is utilised when programs must be downloaded after the cross-assembly process.

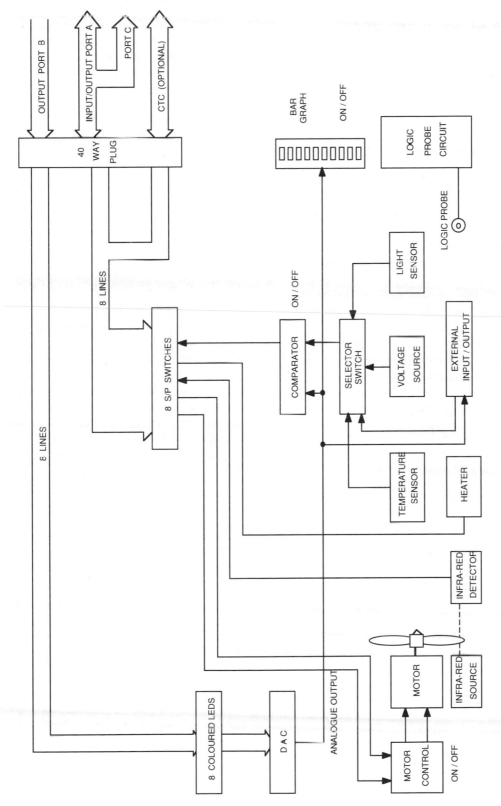

Figure 1.2 Applications board block diagram

1.3 THE POLSU REPLACEMENT EPROM SET

The POLSU replacement EPROM set consists of two EPROMs which are used to replace those supplied with the system. When the POLSU EPROMs are installed, the system sign-on message is either:

POLSU or OLS L3

If the Micro-Professor sign-on message is UPF--1, a number of the exercises will not work correctly. In particular, a number of monitor subroutines will be missing, the downloading facilities for the cross-assembler .HEX files will be absent, and the parallel input/output ports will not be correctly initialised on switch-on.

However, with the original EPROMs the main MONITOR routines will still be present and it is still possible to perform a large number of the exercises in this book. Specifically, a port initialisation program will normally have to precede the software listings given in the text.

1.4 FEATURES OF THE APPLICATIONS BOARD

The applications board consists of a range of circuits and systems designed for a wide range of microcomputer control exercises. It can be connected to any computer with an uncommitted 16-bit input/output port and contains:

- Eight single pole two way switches.
- Eight coloured lights.
- An eight-bit digital-to-analogue converter (DAC).
- A comparator, to provide an 8-bit analogue-to-digital converter (ADC).
- A 10-segment bar graph display.
- A d.c. motor.
- A three-bladed propeller and infra-red beam.
- A heater.
- A temperature sensor.
- A light intensity sensor.
- A potentiometer.
- A logic probe.
- External analogue and digital connections.

Each section of the board may be switched on and off independently to provide maximum system flexibility.

The relationship between the various features listed above is best illustrated by reference to *Figure 1.2* (page 3) which is a system block diagram. It can be seen that two 8-bit programmable ports are used to connect it to the Micro-Professor system and that it can use the Z80 CTC if fitted.

The operation of each section of the applications board is as follows.

Switches and Lights

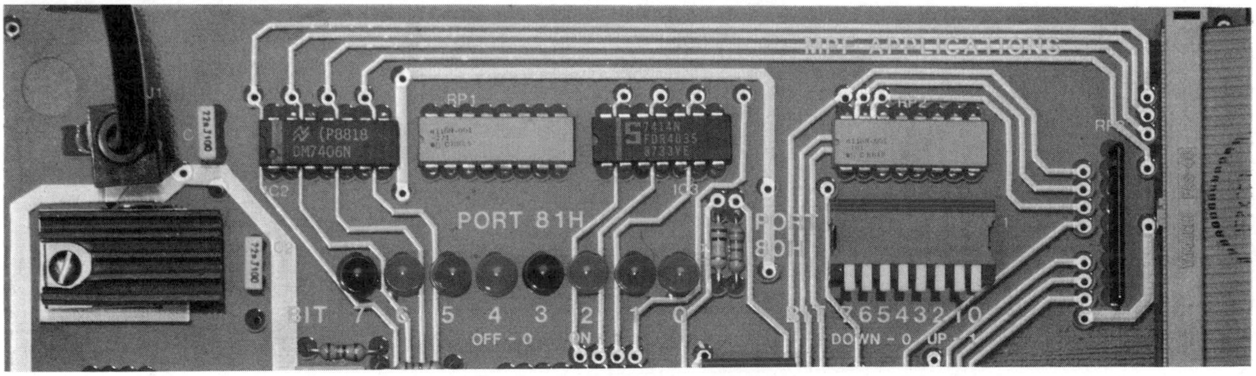

Figure 1.3   Switches and lights

When the **motor, heater, ADC** and **bar graph** switches are all OFF, the board acts as a simple 8 input, 8 output system.

The switches are connected to Port A of the host computer PIO and provide a logic 1 in their UP position and a logic 0 in their DOWN position. Although the switch is small it is very robust and should give reliable operation.

The host system PIO must be initialised as an input on all bits. Since a Z80 PIO is being used then it must be initialised into mode 3, bit control mode which eliminates the need for any handshaking signals.

The switches may also be used as a **manual** control for the **motor** and **heater**.

If the motor switch is turned ON, the motor may be controlled by bits 7 and 6 of the input switch. When these bits are both 0 or both 1 the motor stops. If they are different, the motor turns.

Bit 7	Bit 6	Motor action
0	0	Stop
0	1	Reverse
1	0	Forward
1	1	Stop

Similarly the heater can be controlled by the switch bit 5. If the heater switch is turned ON, then when bit 5 of the input switch is a logic 1 the heater turns ON but when it is a logic 0 the heater turns OFF.

Bit 5	Heater action
0	Heat OFF
1	Heat ON

When the PIO Port A connected to the switches is initialised so that bits 5, 6 and 7 are outputs, then the computer system can control the motor and heater directly and the switches are overridden. It is usual to leave the switches in the logic 1 condition in this case.

The eight lights on the board are connected to the 8 bits from Port B of the system PIO which must be initialised for 8 outputs. They come ON when a logic 1 is applied and stay OFF when a logic 0 is applied.

Logic 0 – Light OFF
Logic 1 – Light ON

Since these lights have different colours it is possible to perform reasonably realistic 'traffic light' sequences etc. In addition, the lights give a visual check of their binary data being output by the port. This is particularly useful when performing digital to analogue conversion experiments since it represents the binary data being converted to an analogue voltage.

Analogue Output and Bar Graph

The output Port B is connected via the 8 lights directly to an 8-bit digital-to-analogue converter (DAC) (*Figure 1.4*). This is arranged to generate an

Figure 1.4 Analogue output and bar graph

analogue voltage in the range 0.00–2.55 volts. A change in the least significant bit produces a voltage change of 10 mV at the analogue output.

The analogue voltage produced by the DAC goes to three other places on the board:

(a) The analogue output external connector on the lower edge of the board.
(b) A voltage comparator which allows the system to be used as an analogue-to-digital converter when suitable control software is written.
(c) The bar graph via an ON/OFF switch.

When the bar graph control switch is ON, the ten segments are illuminated according to the value of the analogue voltage supplied. Segments are illuminated from the bottom upwards, each one corresponding to an analogue voltage change of about 0.24 volts.

The bar graph provides a means of checking the analogue output quickly and can be very useful for monitoring waveforms or voltage sequences produced by the system. When not required simply switch the display OFF.

Analogue Inputs and Heater

Analogue-to-digital conversion must be performed by the system software using the hardware provided on the applications board. This is shown in *Figure 1.5*.

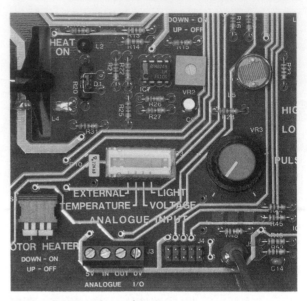

Figure 1.6   Analogue inputs

The analogue voltage produced by the on-board DAC is compared in the comparator with the unknown input voltage. The comparator output is connected directly to bit 3 of the input port A:

- If DAC output is too low – Bit 3 is a logic 1.
- If DAC output is too high – Bit 3 is a logic 0.

To use the analogue-to-digital conversion facility it is necessary to keep the bit 3 switch of the input port at a logic 1.

The comparator output is also connected via the 40-way cable to channel 1 trigger input of the CTC chip (if fitted). The analogue input to the comparator can be derived via a selector switch from one of four sources (*Figure 1.6*). These are:

(a) External voltage source.
(b) Temperature sensor and circuit.
(c) Voltage source.
(d) Light sensor and circuit.

Simply slide the selector switch to the right or left to select the required switch.

An **external voltage** between 0.0 and 2.55 volts may be applied via the external connector near the bottom edge of the board. Great care should be taken to ensure that the external voltage does **not** exceed 5.0 volts.

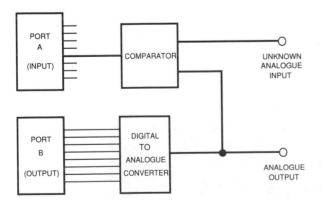

Figure 1.5   Analogue-to-digital conversion system

The **temperature sensor** is a semiconductor diode which is mounted on the surface of the **heater resistor**. The associated circuit produces a linear voltage proportional to the sensor temperature. It is arranged to generate a voltage which changes by 20 mV/°C, starting at about 0.0 volts for 0°C.

The **voltage source** is a simple potentiometer fitted with a control knob. This generates an analogue voltage in the range of 0.0–2.55 volts and may be used in many types of experiment.

The **light sensor** is a light-dependent resistor which responds to a wide range of light intensities. It generates a voltage which rises as the light intensity increases but is not linear. The voltage produced is in the range of 0.02 volts to about 1.8 volts.

The **heater** is not an analogue input, but it can be used to directly control the temperature sensor and thus provide a means of experimenting with closed-loop temperature control. When the **heater** switch is ON, the heater is controlled either by the computer, or by the position of the bit 5 switch on the input port.

THE HEATER GETS VERY HOT – DO NOT TOUCH IT.

The heater surface temperature can rise to about 75 °C so it should not be touched!

When the heater is ON the red light next to it will also be illuminated as a visual check.

Note that the temperature sensor can also be cooled by the **fan** connected to the motor shaft.

Motor Control and Feedback

Bits 6 and 7 of port A are used to control the small d.c. motor which can be made to go in either forward or reverse directions. The **motor control switch** must be ON before the motor will run. It is capable of running at speeds up to about 12 000 r.p.m. and its speed can be controlled by driving it with a pulse waveform if required. The mark to space ratio of such a waveform gives directly proportional speed control.

A three-bladed propeller is connected to the motor shaft (*Figure 1.7*). This acts as a fan for the temperature sensor and it also cuts an infra-red beam as it revolves. As each blade passes through

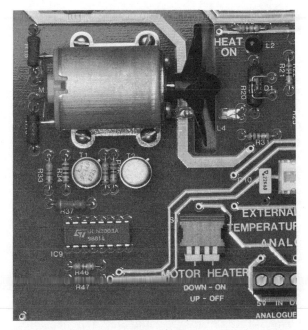

Figure 1.7 Motor control circuits

the infra-red beam a negative going pulse is sent to bit 4 of the input port. This allows revolution counting and timing programs to be developed. The output from the infra-red detector is also sent to the counter timer channel 0 trigger input if fitted.

Logic Probe

An on-board logic probe (*Figure 1.8*) (overleaf) eliminates the need to purchase a separate piece of equipment for fault diagnosis and testing. It provides an indication of HIGH and LOW TTL compatible logic levels as well as a **pulse** light which detects changes in state as short as a few nanoseconds. By interpretation of the brightness of each of the lights it is possible to obtain a reasonable idea of the type of waveform being investigated:

HIGH light ON – 2.4–5.0 volts, Logic 1
LOW light ON   – 0.0–0.8 volts, Logic 0
Both lights OFF – 0.8–2.4 volts,
                          bad level or tri-state

To use the probe, simply connect the lead into the probe socket on the board. It is arranged to allow the testing of waveforms in all parts of the circuit.

Figure 1.8 Logic probe

The probe is particularly sensitive and may indicate **pulse** activity when being held above the circuit. This is due to inductive pick-up and can be eliminated if the operator touches the right-hand pins of connector J4 near the probe socket. When the probe is connected to a circuit point, the indication is not affected by this pick-up.

PRACTICAL EXERCISE 1.1 – DEMONSTRATION PROGRAMS

Experimental Concepts

The first experiment has been designed to introduce the Micro-Professor and the applications board by running some of the demonstration programs which are resident in the computer memory. These programs give a glimpse of the type of applications to which a small system may be put. It will also allow the hardware to be checked for correct operation.

Each program is held in the computer memory at a unique location, known as its **address**. When the computer is given the start address of a program it can then proceed to execute it.

Throughout the Practical Exercise, the following convention has been used. All keys that must

be pressed are enclosed in square brackets, for example [ADDR] indicates the address key. The loudspeaker 'bleeps' for each key depression apart from the four keys in the extreme left-hand column of the keyboard.

Procedure

For the first experiment simply follow the instructions below which will show how to run a program:

(a) Connect up the Micro-Professor, applications board and power supplies if this has not already been done.

To do this, switch off the power to the main board and connect the plug on the ribbon cable to the connector labelled PIO.CTC I/0 BUS, nearest to the display. Check that all the pins are aligned correctly. It is very easy to misplace this connector, so that only one row of pins is inserted, or it is placed too far to the left or right. Double check the connection.

Switch on, and verify that the correct sign-on message appears. It should say OLS – L3 or POLSU.

If it says UPF--1, none of the following exercises will work. Obtain a replacement EPROM set from POLSU before proceeding.

(b) Press [ADDR].

The four left-hand digits on the display now have dots beside them indicating that the computer is waiting for an **address** to be entered. This will be the start address of a program.

(c) Press [2] [0] [D] [9].

This number will step across the displays from right to left as it is entered. The two right-hand displays should now show **dd**.

(d) Press [GO].

This starts the **stopwatch** program which makes the computer operate as a reasonably accurate stopwatch with 1/100 second accuracy.

Press any of the keys on the keyboard apart from those in the left-hand column to stop the display. Release the key to restart it.

The timing accuracy of the stopwatch is obtained because the computer operates at a known speed which is controlled by a quartz crystal. It is the small silver or purple coloured object just above U11 on the Micro-Professor board.

(e) Press [RESET] to stop the program.

(f) The next program will use the small motor on the applications board. Before running the program the motor circuits must be switched on.

Locate the switch marked MOTOR near the bottom left-hand corner of the applications board and make sure it is in the ON position (down).

(g) Press [ADDR] [2] [1] [0] [1] [GO].

The motor should now turn for about 1.5 seconds in alternate directions stopping briefly each time it reverses. Motor control is affected by the computer simply by outputting two control signals to the motor control circuit located just below the motor.

When it has run long enough, press [RESET] to stop it.

After pressing [RESET], the motor can still be controlled **manually**. To do this, put down either bit 7 or bit 6 (but not both) of the 8 pole switch designated port 80H, near the 40 way connecting lead in the top right-hand corner of the applications board.

After finishing with the motor, switch the MOTOR control switch OFF (up).

(h) One of the other features of the applications board is a **bar graph** display. This is connected so that it gives an indication of the analogue voltage being generated by a digital-to-analogue converter on the board.

Switch on the bar graph by putting the switch marked BAR GRAPH into the ON (down) position.

Switch on the analogue-to-digital converter by putting the ADC control switch into the ON (down) position.

Both these switches are located near the centre of the applications board.

Make sure that all the port 80H switches are in the logic '1' (up) position.

Switch the **analogue input** switch to the **voltage** position, number 3. To do this simply lift the right-hand side of the clear plastic cover which is hinged, and slide the switch until the arrow aligns with the required position.

(i) A program that demonstrates the use of the bar graph is located at address 211FH. Press the appropriate keys to run it. There is no H key. That simply means that the address 211F is in hexadecimal.

The bar graph display lights should be rising and falling at a constant rate. Port 81H lights will also flash.

Now turn the knob on the applications board. What happens? The computer senses the analogue voltage generated by the potentiometer and uses it to control the bar graph display.

(j) Press [RESET] to stop the program.

(k) The final demonstration program involves the temperature sensor.

Slide the analogue input selection switch to the **temperature** position, number 2.

Now run the program located at address 2151H.

This program shows the temperature of the temperature sensor on the computer display, in degrees centigrade.

(l) Turn on the heater by putting the heater control switch into the ON position (down). The **heat on** light should come on and the value of temperature indicated on the display should rise.

Do not touch the heater. It can become very hot.

(m) Switch OFF the heater and turn ON the motor control switch.

Now turn on the fan manually by putting either bit 6 or bit 7 of the port 80H switch into the logic '0' position (down). This cools the temperature sensor and the display should respond accordingly.

Note that when the system is reset, the heater can be controlled manually by the bit 5 switch of the 8 position port 80H switch, but of course the temperature will not be displayed.

(n) Finally, press [RESET] to stop the program and also switch OFF the motor, heater and ADC control switches.

(o) The computer software includes a number of useful routines for those students who have difficulty in converting numbers between binary, decimal and hexadecimal. Try each of the following conversion programs with a range of numbers.

Binary to hexadecimal: Start address 0B62H.
Binary numbers entered on the 8-pole input switch are displayed in hexadecimal on the 7-segment display.

Binary to decimal: Start address 21A6H.
Both hexadecimal and decimal numbers equivalent to the binary on the input switches are displayed.

Hexadecimal to binary: Start address 0B3CH.
Hexadecimal numbers entered (in pairs) on the keyboard are shown on the display and also displayed in binary on the eight output lights.

Summary

Programs located at different addresses in the microcomputer memory may be run by pressing the keys:

[ADDR] [START ADDRESS (4 keys)] [GO]

Each program can be stopped by pressing [RESET].

On the application board each part of the circuit has its own **control** switch which must be in the ON position before the chosen circuit will function. In addition, the motor may be controlled manually with bits 6 and 7 of the port 80H switch, and the heater may be controlled with the bit 5 switch.

A slide switch is provided which allows one of four analogue input signals to be selected.

PRACTICAL EXERCISE 1.2 – ENTERING AND RUNNING A SIMPLE PROGRAM

Experimental Concepts

This experiment introduces the method of entering, running and then modifying programs in the computer's changeable memory, known as RAM.

The programs to be entered are designed to show how the microprocessor can read information from the switches on the applications board, and control the operation of the lights.

Procedure

To perform this experiment, the applications board must be connected to the main Micro-Professor board.

(A) Entering a program   The procedure for entering a program must be followed carefully. Similar procedures will be required many times throughout the book so they should become very familiar.

(a) Press [ADDR] [1] [8] [0] [0]
This address 1800, represents the start of user memory (RAM) where most programs will be entered. The four left-hand displays should show 1.8.0.0.

(b) Press [DATA]
The dots have now moved to the two right-hand displays, indicating that the next keyboard entry will be treated as data. The system is now waiting for the program.

(c) Press [D] [B]. This is the first byte of **data**.
As these keys are pressed the display should show them as d and b in the two right-hand indicators. This represents the code for the first instruction of the program.

(d) Press [+]
The display now changes to show the next address 1801 and the two dots indicate that more data is expected.
Continue by entering the **data** from the program below, starting with [8] [0], and following each pair of digits with a [+]. At each step the address provided by the computer should correspond with that in the address column below.

Program 1

Address	Data	Instruction
1800	DB	IN A,(80H)
1801	80	
1802	00	NOP
1803	00	NOP
1804	D3	OUT (81H),A
1805	81	
1806	C3	JP 1800H
1807	00	
1808	18	

The name of each instruction code is given for reference only.

Note that when entering an address at the beginning of a program, or the data, the numbers in the display will simply shift from right to left as more keys are pressed on the keyboard. If a wrong key is pressed, simply continue to press the correct keys until the data in the display is what was originally intended. Then proceed to the next step.

(e) When all the data of the program has been entered, it is a good idea to check it! Return to the start of the program by pressing:

[ADDR] [1] [8] [0] [0]

or by pressing [−] a number of times.

Step through the program slowly by pressing [+] and check at each address that the data corresponds with what is written in the program. If any data values are incorrect, simply press [DATA] and enter the correct values, then continue to press [+] until all the program has been checked.

(B) Running the program

(a) The program is now in the computer memory. To run it press [ADDR] [1] [8] [0] [0] [GO].

(b) Move some of the switches on the applications board and the corresponding light should change.

This is because the computer is reading the data from the switch and outputting the same data to the lights. There is no direct electrical connection between the switches and the lights so that if the program is stopped by pressing [RESET], the switches no longer affect the lights. This point can be further illustrated by modifying the process that the computer carries out on the switch input data.

(C) Modifying the program

(a) Change the program slightly by modifying the data in the addresses shown below:

1802   EE
1803   FF

Proceed as directed in step (e) above, treating the data already in the computer memory as 'incorrect' and replacing it with the new data.

(b) Run the modified program then write down the answers to the following questions.

——————— **Questions** ———————

1.1 How do the switches now affect the lights?

1.2 The input and output can be represented in binary notation as shown below:

Switch up = 1
Switch down = 0
Light on = 1
Light off = 0

Write down the binary output on the lights for the following inputs on the switches. The first digit corresponds with the left-hand switch:

(a) 1 1 1 1 0 0 0 0
(b) 0 1 0 1 0 1 0 1
(c) 0 0 0 0 0 0 0 0

(c) Modify the data again in the addresses shown below:

1802 0F
1803 00

(d) Run the program and answer the following questions:

——————— **Questions** ———————

1.3 How do the switches now affect the lights?

1.4 What is the binary output on the lights for the following inputs on the switches?

(a) 0 0 0 0 0 0 0 1
(b) 1 0 0 0 0 0 0 0

(e) Repeat the last two steps with the new modified data:

1802 F6
1803 AA

Questions

1.5 How do the switches now affect the lights?

1.6 What is the binary output on the lights for the following inputs on the switches?
  (a) 0 0 0 0 0 0 0 0
  (b) 0 1 0 1 0 1 0 1
  (c) 1 1 1 1 1 1 1 1

Summary

This exercise has covered the entry and running of simple programs to operate the input and output of the microprocessor system. The switches and lights used to indicate input and output can be conveniently represented in binary notation since each device has only two possible states, on and off.

The function of the keys used can be summarised as follows:

[RESET] – Reset the system to its initial state.

[ADDR] – Indicate to the system that the next group of keys pressed will be an address, i.e. a place in the computer memory.

[DATA] – Indicate to the system that the next group of keys pressed will represent data, i.e. instructions to be put in the computer memory.

[+] – Add one to the address value.

[–] – Subtract one from the address value.

[GO] – Start to run a program.

[0] to [F] – Numeric keys for addresses and data.

PRACTICAL EXERCISE 1.3 – SINGLE STEPPING A PROGRAM

Experimental Concepts

The programs used in these exercises are designed to use each part of the microprocessor system: the input port, CPU accumulator, memory and output port. It therefore needs the applications board to be connected to the main microcomputer board, as in the previous experiments.

(A) Single stepping example

It is intended that the following program is only operated in the single-step mode i.e. one instruction at a time, which will allow the contents of the accumulator and memory to be examined after each instruction. This will give a clearer understanding of the effect of each program step. If the program is run at full speed, it will be over so quickly that it will apparently have done nothing. Therefore it is imperative that the instructions are followed carefully.

The program will perform two separate tasks:

(a) It will read the data from the input switches and store it in memory location 180B hex.
(b) It will then read the data stored in memory location 0000 hex and display this at the output (lights).

This is illustrated in *Figure 1.9*.

Program

Address	Data	Instruction	Comment
1800	DB	IN A,(80H)	Input the data from
1801	80		the switches
1802	32	LD (180BH),A	Load memory address
1803	0B		180BH with the data
1804	18		from A
1805	3A	LD A,(0000H)	Load data into A
1806	00		from address 0000
1807	00		
1808	D3	OUT (81H),A	Output it to the LEDs
1809	81		
180A	76	HALT	Halt
180B	00	Data	

Procedure

(a) Enter the program starting at address 1800, in the same way as for previous programs.
(b) Check that the program has been entered correctly and then return to the start by pressing:

[ADDR] [1] [8] [0] [0]

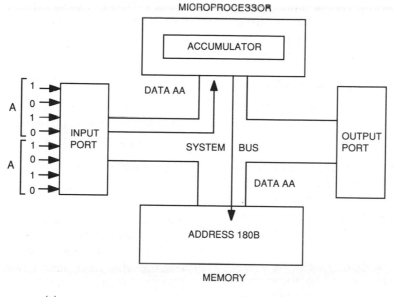

(a)

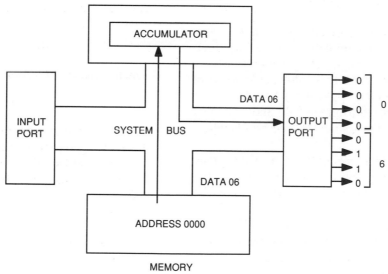

(b)

Figure 1.9 Moving data

(c) Set the input switches on the applications board to the binary value 1 0 1 0 1 0 1 0. This is equivalent to AA in hexadecimal, which is the number system used for the seven-segment display.

(d) Press [STEP]. This causes the first instruction in the program to be executed, and the display shows:

1802 32

which indicates that it is ready to perform the next instruction in the program. However, before proceeding with the next one, examine the effect of the first input instruction on the accumulator in the microprocessor, as follows.

(e) Press [REG]. The display shows rEg-, indicating that it is ready to display the register contents requested by the next key. The names for the registers are given **above** the numeric (white) keys on the keyboard.

(f) Press [AF] (which is the same as pressing [0]). The display shows:

AA80 AF

The two left-most digits represent the contents of the accumulator, which has been loaded with the value AA by the first instruction in the program. This value was input from the switch on the applications board.
Note. Take care to press [AF] and not [AF′] (above the 4 key), since AF′ represents different registers.
The two digits in the centre of the display may not be 80! They represent the contents of the **flag** register and can be anything. For the moment, just ignore them!
The two right-hand digits on the display AF indicate that the display is showing the contents of the accumulator and flags.

(g) Now return to where the program was left by pressing:

[PC]

PC stands for program counter, and it is used to keep track of where a program is up to at any moment. The display shows:

1802 32

as before.

(h) Press [STEP] to execute the next instruction. The display now shows:

1805 3A

The instruction just executed should have put the contents of the accumulator (AA) into memory address 180B, which previously had 00 in it. To examine memory address 180B,

Press [ADDR] [1] [8] [0] [B]

The display now shows:

180B AA

indicating that the accumulator contents have been deposited there by the last instruction.

(i) Return to the program by pressing [PC], then execute the next instruction by pressing [STEP]. The display now shows:

1808 d3

(j) The instruction just executed should have loaded the accumulator from memory address 0000. This address contains the number 06. Check that it has been loaded correctly by pressing:

[REG] [AF]

The *two left-hand displays* now show

06

indicating that the accumulator contents have been changed to 06 by reading the value from memory address 0000. Now examine the contents of address 0000, by pressing [ADDR] [0] [0] [0] [0], to check that it does contain 06.

(k) Press [PC] and then [STEP] to execute the next instruction.
The display shows:

180A 76

and the output LEDs on the applications board show the binary number 0 0 0 0 0 1 1 0, which is the binary equivalent of the hexadecimal 06. This number has been sent from the accumulator to the LEDs by the last instruction executed.

(l) Press [GO] to complete the program. The display should blank out, and the red, HALT LED should come on, indicating that the processor has halted. Before proceeding any further, press [RESET] to restart the Z80.

(B) Single-stepping exercise

Introduction This exercise is very similar to the last one but by summarising the effect of each instruction on various parts of the system, it is possible to produce a simple trace table.

The program will first load the accumulator with data from memory address 180B hex and output it to the lights. It will then read the input switches and store the data in address 180B hex.

Program

Address	Data	Instruction	Comment
1800	3A	LD A,(180BH)	Load accumulator
1801	0B		from address 180BH
1802	18		
1803	D3	OUT (81H),A	Output A contents
1804	81		to the LEDs
1805	DB	IN A,(80H)	Input data from the
1806	80		switches
1807	32	LD (180BH),A	Store the data in
1808	0B		address 180B
1809	18		
180A	76	HALT	Halt
180B	0F	Data	Initial data

Procedure

(a) Enter the program into the memory in the usual way and check it.
(b) Set the input switches to binary 0 1 1 1 0 1 1 1 (77 in hexadecimal).
(c) Return to the start of the program.
(d) Step through **one** instruction. Examine the contents of the accumulator, memory address 180B and the output LEDs. At the same time enter the data in the table.
(e) Repeat Step 4 until the program is completed. The table below should be completed with great care. This type of table is a simplified version of a 'trace table', which is dealt with in greater detail in a later exercise.

Step	Address	Accumulator	Memory address 180B	Output LEDs (binary)
0	1800	X	0F	0 0 0 0 0 0 0 0
1				
2				
3				
4				
5				

X = Unknown – could be anything!

Summary

In this exercise stepping through a program one instruction at a time has been examined. It is possible to see how data is transferred around the system between input port, accumulator, memory and output port. All data has been transferred via the data bus; the address bus has been used to indicate where data must come from or go to and the control bus has been used to activate each part of the system at the right time.

The function of the new keys used can be summarised as:

[STEP] – Executes one instruction in a program when the instruction starting address is showing in the four leftmost displays.

[REG] – Allows the user to examine the contents of registers in the microprocessor.

[AF] – Indicates the contents of the Accumulator in the two left-hand displays. (Only when used after [REG].)

[PC] – Returns to the point in the program where it was stopped.

After each **step** of the program, the accumulator contents may be examined by pressing:

[REG] [AF]

Also, any memory address may be examined by pressing:

[ADDR] [Required Address]

It is then possible to return to the program by pressing:

[PC]

The exercise has shown how data can be transferred from one part of the microprocessor system to another. A simple trace table is a record of the effect of each instruction on the system. Trace tables provide a good way to see how data moves from one part of the system to another.

PRACTICAL EXERCISE 1.4 – SYSTEM MEMORY MAP

Experimental Concepts

The memory map of a microprocessor system gives a detailed record of the addresses of different types of memory within the system. In its simplest form it indicates which addresses are ROM, which are RAM and which are unused. More precise memory maps indicate the allocation of different areas of each type of memory for special functions. For example, the Micro-Professor has a number of different programs in ROM. A detailed memory map would indicate where each one of these is. Spaces in RAM that are reserved for special functions are also often shown on memory maps.

In this exercise the range of addresses for different types of memory may be worked out within the microcomputer system and hence a simple memory map plotted.

Remember that memory chips come in specific sizes, generally multiples of 1K, i.e. 1024 bytes. Therefore it is impossible to have say, 10 bytes of ROM followed by 50 bytes of RAM. When checking the memory map, it is not necessary to test every byte of memory, but simply the '1K boundaries'. A 1K memory chip starting at address 0000 hex would occupy addresses 0000 hex to 03FF hex and 0400 hex is said to be the first 1K boundary.

Note that 0400 hex = 1024 decimal.

Each 1K block of memory could consist of ROM, RAM or no memory at all. RAM can be changed by entering data from the keyboard in the normal way. ROM contains data but cannot be changed.

Trying to change the ROM contents from the keyboard will result in the display going BLANK and returning to the previous value when the key is released. If no memory is present the display shows FF for the data, which cannot be changed. Note that a ROM could also contain FF which cannot be changed, but not usually on a 1K boundary and certainly not in the Micro-Professor system.

Procedure

It would take a very long time to investigate the full 64K address range of the Z80 microprocessor, so in the experiment our attention is limited to the first 16K.

(a) Start by working out all of the 1K boundary addresses for the first 16K of memory in hexadecimal.

Draw a chart similar to that shown in *Figure 1.10*, with all 16 of the addresses on the left-hand side.

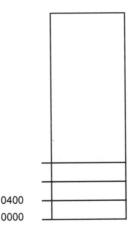

Figure 1.10 Memory map

(b) Now test each 1K boundary address and mark on the chart the type of memory found in the 1K above this address.

(c) Since there are only three memory chips in the system, write down the size (in bytes) of each chip. State which chip corresponds to each memory block on the diagram.

(d) A simple RAM test program exists at address 069A. Before running it:

(i) Load the HL register pair with the start address of the area to be tested.

(ii) Load the BC register pair with the number of bytes to be tested.

When the program is run, if all the bytes tested are RAM, the display will show OLS–L3, or POLSU.

If any byte is not RAM, the HALT light, to the right of the display, will come on. When this happens press [MONI] [REG] [HL] to display the faulty address, then [ADDR] to get the content of that byte.

Run the program a few times and experiment with different values in HL and BC. At least verify the addresses of the RAM space on the memory map.

Summary

A memory map is useful both to system designers and to programmers since it gives an indication of the type of memory found at different addresses in a system. Simple tests, operated from the keyboard, allow users to determine the type of memory found at each 1K boundary in a system. This then allows a simple memory map to be constructed.

Memory test programs are available which check each address within a given range and indicate whenever the byte found is not RAM.

TWO

Tracing program operation

EQUIPMENT REQUIRED

To complete all of the practical exercises in this chapter you will need the following:

(a) *Micro-Professor MPF-1B.*
(b) *Applications board MAB with logic probe lead.*
(c) *The POLSU replacement EPROM set.*
(d) *The Micro-Professor single step board.*

PREREQUISITES

Before studying this chapter you should have:

(a) *An understanding of the various parts of a microcomputer system.*
(b) *An appreciation of the Z80 fetch–execute cycle.*
(c) *An awareness of a few simple Z80 machine code instructions in the data transfer group.*
(d) *An appreciation of typical instruction timing operations.*

These may be obtained by studying the first two chapters in the companion book in the series, *Microelectronics NII*, together with the first part of Chapter 3 which deals with the data transfer instructions.

2.1 INTRODUCTION

This chapter builds on the work of the previous one by exploring the actual operation of machine code programs in a detailed way. It is vital that students who wish to understand how the computer works rather than simply appreciating what it does, fully grasp the basic concepts of bus activity and timing.

The equipment required to examine the operation of computers at normal speed, such as logic analysers, are very expensive and assume a lot of previous knowledge. The exercises in this chapter use nothing more complicated than a logic probe and a special single-step board for the Micro-Professor to allow bus activity to be examined and understood. By relating the information on the logic probe or single-step board to the program being executed it is possible to identify the activity related to each instruction and its timing waveform.

Some of the facilities of single board microcomputers are also useful in their own right as debugging aids. For example, the use of a single-step key and a breakpoint facility allow students to debug their own programs when they go wrong, as they inevitably will. Therefore it will be useful for the more complicated programs in future to be able to utilise all of the debugging aids available with the system.

PRACTICAL EXERCISE 2.1 – USING THE LOGIC PROBE

Experimental Concepts

Signals within a microcomputer system can have one of three states:

0.0–0.8 V: logic 0 – LOW
0.8–2.4 V: bad logic level (tristate)
2.4–5.0 V: logic 1 – HIGH

A logic probe is designed to indicate clearly these three states and also to show any very rapid transitions between them. By interpreting the indications shown on the logic probe lights it is possible to deduce the type of waveform being examined.

The three logic probe lights on the applications board are labelled **high**, **low** and **pulse**. The high light indicates a high voltage (above 2.4 V) which corresponds with a logic 1 state. The low light indicates a low voltage (below 0.8 V) which corresponds with a logic 0 state. Note that high and low have been used for the lights rather than 1 and 0 since, in some circumstances, it is desirable to reverse their logic states. This is known as negative logic. However, unless otherwise stated, in this book high will indicate a logic 1 and low a logic 0.

The pulse light flashes whenever there is a rapid change from one logic state to the other.

It is very important to be able to connect the type of waveform being examined, with the type of indication shown on the lights. This can be explained best by examining *Figure 2.1* (page 20).

In general terms the brightness of the high or low lights gives an indication of the time that the waveform spends in each state. Therefore, some rough estimation of the mark to space ratio (logic 1 to logic 0 ratio) of the waveform can be obtained by examining the brightness of the lights. The pulse light flashes at the same rate no matter how rapid the pulse train, once there are more than about three pulses per second. Below that pulse rate, one flash represents one change of logic state.

Many of the waveforms to be examined are associated with the CPU buses. *Figure 2.2* (page 21) shows the pin connections of the Z80. The most important pins for the moment are the address lines A_0 to A_{15}, the data lines D_0 to D_7 and the system control lines which form part of the control bus. The Z80 CPU is designated U1 on the Micro-Professor board and is the chip in the top left-hand corner.

When using the logic probe it is important to make a positive connection with each pin but not so firm that there is a danger of making the probe slip off to the side. The *best place* to connect to each pin is at the point where the pin enters the socket. Here, there is a small recess which can accept the probe point safely.

Procedure

(A) Checking the logic probe The first part of the experiment is to check the operation of the logic probe by examining some known waveforms in the system:

(a) Ensure the applications board is connected to the Micro-Professor. Connect and switch on the mains power supplies to both boards. The sign-on message should appear on the display.

(b) Connect the logic probe lead to the socket on the applications board.

(c) Quickly check the **high** and **low** lights by putting the probe on the 5 V and 0 V screw heads on the analogue input/output connector. The lights should show high and low respectively.

(d) Now check the power supply pins of the Z80, pins 11 and 29. Ensure that they give the expected results.
   Note that pin 1 is in the top left-hand corner of the integrated circuit and pin 21 in the bottom right corner.

(e) Now check pin 6. This is the **clock** input to the Z80. Both high and low should come on (one may be slightly brighter than the other) and the pulse light should flash.
   The waveform here approximates to a high frequency squarewave.

(f) Although the computer does not appear to be doing anything, it is actually executing its **monitor** program. This program continuously checks the keyboard waiting for a key to be pressed and keeps the display updated.
   Check all of the address and data pins on the Z80. There should be considerable 'bus activity' with most lines indicating some pulse waveform.

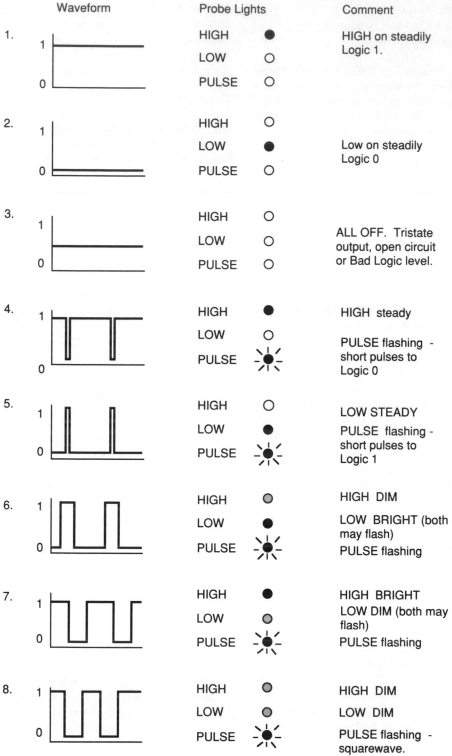

Figure 2.1   Logic probe interpretation

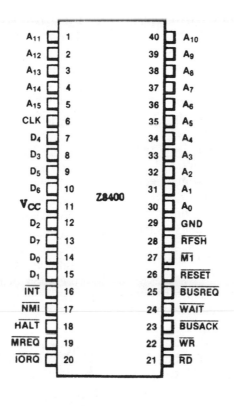

Figure 2.2 Z80 pin connections

──────── Question ────────

2.1 Which address bus lines do not show any pulse activity when the sign-on message is showing on the display?

(g) Finally check pin 26, the **reset** input. Note what happens when the [RESET] key is pressed on the keyboard.

(B) Programmed activity All activity on the address, data and control buses in the microcomputer is dependent upon the program that the system is running. For example, if the only addresses used in a program were 1800, 1801 and 1802, then some of the address lines (e.g. A_{13}, A_{14} and A_{15}) would be expected to remain at logic 0 all the time the program was running.

The control signals generated also depend upon the instructions used. In the Z80, there are four main system control signals:

$\overline{MREQ}$ – Memory request: used to access memory.

$\overline{IORQ}$ – Input/output request: used to access ports.

$\overline{RD}$ – Read: for all read operations.

$\overline{WR}$ – Write: for all write operations.

The bars over the signals indicate that they are **'active low'**. This means that their normal state is a logic 1, but whenever say, a **read** operation takes place the $\overline{RD}$ line goes to a logic 0.

If a program contains no instructions which require a **write** operation, then the $\overline{WR}$ control line should remain at logic 1 all the time.

Program 1

Address	Hex code	Instruction
1800	C3	JP 1800H
1801	00	
1802	18	–

(a) Program 1 has only one instruction which jumps to itself. Write out the addresses and hex codes in **binary** showing clearly the state of each address and data line for each byte.

Note which address and data lines are at the same logic state for all bytes, which should therefore generate no bus activity.

(b) Enter the program and execute it.

(c) Now make a list of all the address and data lines as well as MREQ, IORQ, RD and WR.

(d) Use the logic probe to check the state of each of the lines listed above and write down the type of activity found beside each one in the list. Does this correspond with that expected?

Any discrepancies may be explained by the following facts:

● Address bus line A_0 to A_6 are used for other addresses which **refresh** dynamic memories (if fitted) between instructions.

● Address lines 'float' between the T-states in which they are active. This means that their level is likely to be between a 1 and a 0.

● Data lines are pulled high between the T-states in which they are active.

───────── Question ─────────

2.2 Explain the activity found on the control lines listed above while the program is running.

(e) Now run the program at address 23B7H, which should make the lights on the application board flash.

Answer the following questions.

───────── Questions ─────────

2.3 Explain the activity found on the control lines when the program is running. In particular explain the significance of the flash rate on the WR and IORQ lines.

2.4 Explain the activity found on pins 27 to 34 of the programmable input/output chip (U1O) and hence deduce their function.

(f) If time permits, investigate some of the signals found in other parts of the system.

Write some simple programs to force certain things to occur. For example, what happens on the buses when a **halt instruction** is executed? What activity occurs when the [RESET] key is held down?

Summary

With practice, a logic probe can be used to provide considerable information on the activities within a computer system. In its simple state it will indicate the presence of logic 1s or 0s, but with the addition of a **pulse** light and with correct interpretation of the other lights, many types of waveform can be discovered.

PRACTICAL EXERCISE 2.2 – PRODUCING TRACE TABLES

Experimental Concepts

A trace table is a means of showing the effect of the instructions in a program on the CPU registers and sometimes on memory locations or input/output ports. It is a good way of checking that a program is working correctly. If the contents of registers or memory locations are to be examined during program excecution it is not possible to run the program at full speed. Instead the program is run one **step** at a time and after each instruction step the contents of registers or memory locations can be examined.

The Micro-Professor provides a number of keys which facilitate the production of trace tables. For example, the [STEP] key executes a program one instruction at a time. The [REG] key allows the contents of the registers to be examined, and the [PC] key allows the operator to return to the point which had been reached in the program before the registers or memory locations were examined.

This exercise involves the production of a number of trace tables. The programs have not been reproduced in their assembly language form but it will be quite possible to work them out by writing down the mnemonics from the machine code program in the computer memory. By pressing the keys stated in the experiment, the computer will load a program into RAM starting at address 1800 hex as though it had been entered from the keyboard. It will also clear the registers.

All that is required is to step through the program and produce the trace table.

Procedure

(a) The first program is at address 23CA hex. It only contains instructions from the 8 and 16 bit load groups.

Run the program by pressing:

[ADDR] [2] [3] [C] [A] [GO]

The display should show SYS–SP.
Now press [RESET].

(b) Press [ADDR] [1] [8] [0] [0] and then [STEP].

Examine the registers and addresses listed on the trace table and complete its second line.

(c) Now press [PC] to return to the point where the program was left and then press [STEP] to execute another instruction.

Write down the register and address contents as before.

(d) Continue by repeating (c) until no changes occur on the trace table for two consecutive steps. The trace table should be presented as shown below:

Registers								Memory		
PC	A	F	B C D E H L					1830	1831	1832
1800	00	00	00 00 00 00 00 00					00	00	00
1802										

(e) Now try the same thing again. This time the program to run is located at address 23F5 hex.

Run this program, press [RESET], then step through the program at address 1800H and produce a trace table similar to that above.

Once again the instructions in the program have been limited to those from the **load** group.

Now try the following questions.

2.5 Without entering the program below into the computer, write down the trace table that would be obtained when the program was executed. Assume that all registers and memory locations contain 00 initially.

```
START:  LD HL,1B00H
        LD (HL),25H
        LD B,(HL)
        LD A,B
        LD DE,1B20H
        ADD A,50H
        INC B
        LD (DE),A
        LD (HL),B
        HALT
```

2.6 Write the program that contains only load instructions which generated the trace table below. All registers and addresses initially contained 00 hex.

Registers								Memory	
PC	A	B	C	D	E	H	L	1830	1831
1800	00	00	00	00	00	00	00	00	00
1802	00	00	17	00	00	00	00	00	00
1804	00	78	17	00	00	00	00	00	00
1807	00	78	17	00	00	18	31	00	00
1808	00	78	17	00	00	18	31	00	78
1809	17	78	17	00	00	18	31	00	78
180C	17	78	17	18	30	18	31	00	78
180D	17	78	17	18	30	18	31	17	78

Summary

A trace table is a means of representing the contents of registers and memory addresses during the execution of a computer program. It allows the user to check for the correct program operation by comparing what is expected to happen with what actually happens.

PRACTICAL EXERCISE 2.3 – USE OF BREAKPOINTS

Experimental Concepts

When a program does not work correctly the programmer has a number of alternatives in attempting to remedy the problem. The first is to make sure that the code actually entered into the machine is the same as that which was intended. The second is simply to check the program on paper and attempt to find a programming fault. If both of these checks fail to find the problem it is then possible to **step** through the program one instruction at a time and examine registers and memory to find out if the program is performing as expected. In a long program this can become very tedious and is relatively time consuming. Another alternative is to set a **breakpoint**.

The breakpoint is a means of stopping a program in a controlled way so that registers and memory can be examined at a known point in the program. This eliminates the need to step through the program which can be very time consuming especially in programs containing loops and other lengthy procedures.

The procedure for setting a breakpoint in the Micro-Professor is very straightforward.

When the Micro-Professor display is in its address and data state, i.e. when examining a memory address or entering a program, simply press the [SBR] key to set the breakpoint. Note, however, that [SBR] *must be pressed on the first byte of an instruction, i.e. on an op-code.*

When [SBR] is pressed, all six decimal points will light on the display to indicate the breakpoint. Also when the program is examined later the same thing will happen, so that it is possible to find the breakpoint in a program just by examining the contents of memory.

If a breakpoint has been set the program can then be run in the usual way. However, as soon as the breakpoint address is encountered the program will stop. The display will show the address of the instruction **after** the breakpoint. At this point it is possible to examine the contents of registers or addresses in the normal way.

To continue with the rest of the program after the breakpoint simply press [PC] and [GO]. If the program contains a loop it may stop again when the breakpoint address is reached a second time. It is also possible to **step** through the program after the breakpoint.

If the breakpoint has served its purpose it can be **cleared** by pressing the [CBR] key. When the key is pressed the display shows FFFF FF. This indicates that the breakpoint is set to address FFFF well away from the RAM area. Return to the program by pressing [PC].

The breakpoint does not have to be cleared before changing its address. When a new breakpoint address is required simply display the new address and press the [SBR] key. The old breakpoint will then be ignored.

Program

The following program may be used to illustrate the application of a breakpoint. It is designed to send binary numbers to port 81H but runs very fast so that the individual outputs are difficult to see. However the built-in delay loops make it very tedious to check each individual step.

Address	Hex code	Mnemonic	
1800	06 00	START:	LD B,00
1802	78	LOOP:	LD A,B
1803	D3 81		OUT (81H),A
1805	3C		INC A
1806	47		LD B,A
1807	3E 00		LD A,00H
1809	3D	DEL:	DEC A
180A	C2 09 18		JP NZ,DEL
180D	C3 02 18		JP LOOP

Procedure

(a) Make sure that the applications board is connected to the computer and has power applied.

(b) Enter the program above into the Micro-Professor and run it.

Lights 0–3 on the applications board should appear to be on continuously and lights 4–7 appear to be flashing. However, this is not actually what is happening!

(c) Stop the program then step through it one instruction at a time.

Very soon the display oscillates between addresses 1809 hex and 180A hex. This is because the program incorporates a short delay loop. Examine register A at a number of these steps to confirm that it is counting down.

Unfortunately, the [STEP] button will have to be pressed 512 times before the loop finishes so it is advisable to abandon the stepping.

(d) Instead, set a **breakpoint** at address 180D hex. This should stop the program every time it has completed the instructions once.

(e) Run the program again from address 1800 hex.

When the program stops the lights on the applications board should all be off and the display should show 1802 hex in the address space.

(f) Press [GO] to execute the second loop through the program.

It is now possible to see clearly how the program is affecting the lights on the applications board.

(g) Keep pressing [GO] until the program operation is clear.

———— Question ————

2.7 Which other addresses could be used for the breakpoint to obtain results similar to those obtained previously?

(h) Clear the breakpoint and confirm that the program still works as before.

———— Questions ————

2.8 Enter the program below into the Micro-Professor and execute it. By setting suitable breakpoints describe briefly what the program function is.

Address	Hex code
1800	06 80
1802	3E 00
1804	3D
1805	C2 04 18
1808	78
1809	D3 81
180B	07
180C	47
180D	C3 02 18

2.9 The program below is used to produce analogue waveforms.

Address	Hex code	Mnemonic	
1800	06 20	START:	LD B,20H
1802	21 00 19		LD HL,1900H
1805	7E	LOOP:	LD A,(HL)
1806	23		INC HL
1807	D3 81		OUT (81H),A
1809	11 00 10	DELAY:	LD DE,1000H
180C	1B	DEL:	DEC DE
180D	7A		LD A,D
180E	B3		OR E
180F	C2 0C 18		JP NZ,DEL
1812	05		DEC B
1813	C2 05 18		JP NZ,LOOP
1816	C3 00 18		JP START
1900	N1	TABLE:	DATA
1901	N2		DATA
1902	N3		DATA
–	–		DATA
–	–		
191F	N32		DATA

By using the single-step facility and the breakpoint facility, produce a trace table for the program above showing the contents of the CPU registers and output port 81H. Assume that all the registers are initially zero, and that the data in the table is 00, 08, 10, 18, etc.

Notice that with programs of this type the trace table will be very long if every step is included, because of the delay. Simply produce the trace table for the first time through the delay and then assume it is finished. The same applies to the other conditional jumps.

Summary

A **breakpoint** is a means of stopping a computer program in a controlled way at any convenient instruction. It is a function provided as part of the monitor program. When the program being executed reaches the chosen breakpoint it halts and displays the next address in the program. This allows the registers and memory to be examined. A breakpoint may be set by pressing the [SBR] key when the display is showing address and data. It may be cleared with the [CBR] key.

PRACTICAL EXERCISE 2.4 – MACHINE CYCLE STEPPING A PROGRAM

Experimental Concepts

This exercise is similar to the single-stepping exercise carried out in Practical Exercise 1.3, but it goes beyond that exercise since it is now possible to examine the content of the system buses during each machine cycle. This will not only allow the data and addresses to be seen, but it will also allow the 'invisible cycles', which are the 'execute' parts of each instruction, to be seen as well. The exercise will require the connection of the single-step board to the Micro-Professor, so this should be connected at the beginning. To do this, switch off the computer and carefully connect the 40-way socket to the header at the top left-hand corner of the Micro-Professor board using the short ribbon cable provided. Note that the single-step board is attached to the top connector of the Micro-Professor. Also connect the applications board to the lower of the two 40-way connectors on the Micro-Professor board if it is not already connected.

Power should be connected to the single-step board from the power supply in the normal way and from there via the small jumper provided, to the applications board. The arrangement is shown in *Figure 2.3*.

SINGLE STEPPING BOARD

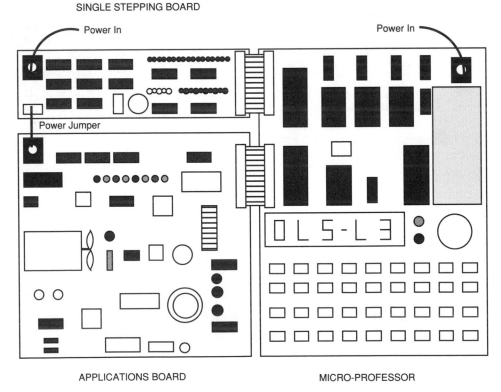

APPLICATIONS BOARD   MICRO-PROFESSOR

Figure 2.3   Connecting the single-step boards

The single-step board provides a means of examining the address bus, the data bus, and the main Z80 control signals. The longest line of LEDs on the board represents the address bus with address line A_{15} on the left-hand side and address line A_0 on the right-hand side. These LEDs have been grouped in fours to make it more convenient to interpret the address bus contents from the binary data. When an LED is ON this represents a logical 1.

The data bus contents are shown on the red LEDs below those of the address bus. These have also been divided in two groups of four to allow easy interpretation of the binary data.

The green LEDs on the board represent the control signals. These are, left to right $\overline{M1}$, $\overline{WR}$, $\overline{RD}$, $\overline{IORQ}$, and $\overline{MREQ}$. The reason for the use of green LEDs here, is that they are the inverse of the actual signals on the control bus, so a light comes ON when the signal is active. On the control bus this represents a logic 0. The lights can be interpreted by reference to *Figure 2.4* where the five main cycles are shown, together with the corresponding lights which would be illuminated.

To the left of the control bus LEDs are a single-step button and a switch to allow normal or single-step operation. When this switch is in the normal position, the whole system can be operated at full speed, but whenever it is placed in this single-step position the system will stop as soon as the [GO] button is pressed on the keyboard. It will stop at whatever address has been set on the Micro-Professor keyboard. This makes operation very simple and is almost the same as the instruction step used previously.

Program

Address	Data	Instruction	Comment
1800	3A	LD A,(180BH)	Load accumulator
1801	0B		from address 180BH
1802	18		
1803	D3	OUT (81H),A	Output A contents
1804	81		to the LEDs
1805	DB	IN A,(80H)	Input data from the
1806	80		switches
1807	32	LD (180BH),A	Store the data in
1808	0B		address 180B
1809	18		
180A	76	HALT	Halt
180B	0F	Data	Initial data

This program is the same as the one used previously, but by using the hardware step facilities much more information can be obtained from the system. In particular, it will be possible to examine how each instruction is executed during the program.

Procedure

(a) First, make sure that the single-step board and the applications board are properly connected to the Micro-Professor and that power is applied to **both** boards as described previously.

Place the **normal/step** switch in the NORMAL position.

(b) Enter the program into the memory in the usual way and check it.

(c) Set the input switches on the applications board to 0 1 1 1 0 1 1 1 (77 hexadecimal).

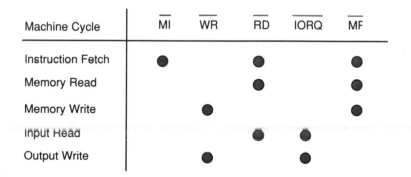

Machine Cycle	$\overline{M1}$	$\overline{WR}$	$\overline{RD}$	$\overline{IORQ}$	$\overline{MF}$
Instruction Fetch	●		●		●
Memory Read			●		●
Memory Write		●			●
Input Read			●	●	
Output Write		●		●	

Figure 2.4 Machine cycle control signal summary

(d) Return to the start of the program.
(e) Put the switch on the single-step board into the STEP position and press [GO] on the Micro-Professor keyboard. The display should now be blank but two lights should be ON, on the single-step board address bus LEDs.

These should represent address 1800 in binary.

In addition the lights on the data bus should represent 3A in binary and the $\overline{M1}$, $\overline{RD}$, and $\overline{MREQ}$ lights on the control bus should be ON.
(f) At this stage it is advisable to begin recording the results observed on the lights and this can be best achieved by completing *Table 2.1*. This shows the step number, the address, data and control bus contents, together with the type of cycle being performed.

Table 2.1

Step	Address bus	Data bus	Control bus	Cycle type
1	1800	3A	$\overline{M1}$ $\overline{RD}$ $\overline{MREQ}$	Instruction fetch
2				
3				
4				
5				
6				
7				
8				
9				
10				
11				
12				
13				
14				
15				
16				

The first entry has been made in the table already.

(g) Now press the STEP button on the single-step board once and observe the lights. These now indicate what is happening during the next machine cycle. For each set of lights translate their contents into hexadecimal as appropriate and enter the results into *Table 2.1*.

Continue to step through the program until it is completed and the HALT light on the Micro-Professor comes on.

2.10 For each of the instructions in the program write down the number of machine cycles required to complete it.

2.11 During the operation of the input and output instructions, the port address appears twice. Where are these?

(h) It is now possible with the information from the single-step exercise to create an accurate timing diagram to show the precise nature of the signals that occur on the system buses. Even though this is a relatively short program with only 16 machine cycles, it would still represent a very complex set of waveforms if the whole program was to be plotted on a waveform diagram. Therefore it will be simpler to represent only one or two instructions in this way.

Figure 2.5 (opposite) shows the waveforms that would be present during the first instruction execution which loads the data 180B hex into the accumulator.

The waveforms are drawn simply by examining the cycles that have taken place during the instruction execution and drawing the corresponding machine cycle from the appropriate waveforms. *Figure 2.5* shows how this can be done.

When the single-step board is used, the mechanism that halts the processor during each machine cycle, stops it at the end of the second T-state. Therefore, whatever data is present on the buses at the point in the cycle is displayed on the lights. A number of actions take place, particularly in the first machine cycle, M1, which cannot be seen even by this method. In particular, the refresh cycle which is a peculiarity of the Z80 microprocessor still remains invisible. During T-state 3 and 4 an address is placed on the address bus which can be used to refresh the contents of dynamic memory devices. At the same time the $\overline{MREQ}$ signal goes low together with the refresh signal. The Micro-Professor does not contain any dynamic memory devices so this part of the cycle can be ignored.

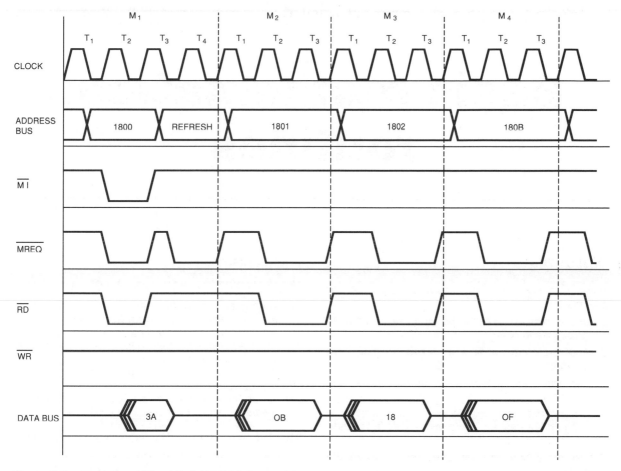

Figure 2.5 Operation of the LD A,(180BH) instruction

─────── **Question** ───────

2.12 From the experimental results, draw wave-form diagrams which correspond to the operation of the following instructions:

(a) IN A,(80H).
(b) LD (180BH),A.

Summary

The use of the single-step board allows the bus activity during each machine cycle of an instruction to be observed. During each machine cycle the information is halted on all the buses at the end of T-state 2 and this allows the machine cycle data to be interpreted. This is particularly important when the CPU introduces machine cycles which are not directly related to reading data from the memory.

THREE

Basic mathematical routines

OBJECTIVES

When you have completed this chapter, you should be able to:

1. *Use a microcomputer to perform basic arithmetic processes in binary.*
2. *Appreciate the significance of **two's** complement notation.*
3. *Perform multiplication and division of binary numbers.*
4. *Understand how large numbers are handled in microcomputers by performing multi-byte arithmetic operations.*
5. *Use the DAA instruction effectively in BCD arithmetic.*

EQUIPMENT REQUIRED

To complete all of the practical exercises in this chapter, you will need:

(a) *Micro-Professor MPF-1B.*
(b) *Applications board MAB.*
(c) *The POLSU replacement EPROM set.*

Since the programs are becoming relatively long in this chapter, it may be best to use a cross assembler on an IBM PC or compatible computer and write the programs in assembly language. In this case, you will also require:

(d) *PC with cross assembler, editor, linker and connecting lead.*
(e) *(Optional) practical exercises on **disk**.*

PREREQUISITES

Before studying this chapter you should have:

(a) *An appreciation of the main Z80 instructions from the **load**, **arithmetic** and **logic** groups.*
(b) *An understanding of the basic concepts of binary numbers including their different representations within a computer system.*
(c) *An appreciation of the function of the **flag** register in a microprocessor.*
(d) *An ability to perform binary mathematics such as inversion, addition and subtraction.*

These may be obtained by studying the first three chapters in the companion book in the series, *Microelectronics NII*. This also introduces the concepts of assembly language. However, if you intend to use a cross-assembler to develop the programs in this chapter, it is recommended that you also study the practical exercise in the appendix to this book which explains the method in detail.

3.1 INTRODUCTION

In view of the highly complex mathematical operations that computers can perform, it is sometimes surprising to realise that most microprocessors can only add and subtract.

The exercises in this chapter explore the concepts of binary arithmetic and show how simple addition and subtraction operations can be used to perform much more complex functions. At the root of all these calculations are binary numbers in two's complement notation and the use of those is first examined. Students are encouraged to check all of the calculations on paper to ensure that the programs are doing exactly what they are supposed to do.

Once the basic operations have been mastered, simple routines for multiplication and division are developed. These employ the 'shift and add or subtract' methods which provide relatively rapid results. They can also be extended with some care to deal with numbers with many more bits.

Much larger numbers and the special routines needed to manipulate them are also covered in one of the exercises. Multi-byte addition and subtraction operations are examined with the help of some basic algorithms. This is extended to the use of binary coded decimal format for numbers which must be displayed in the final practical exercise. In doing so, the operation of the decimal adjust accumulator instruction is examined in detail.

PRACTICAL EXERCISE 3.1 – BINARY ARITHMETIC

Experimental Concepts

This exercise is designed to show how some of the arithmetic operations of the Z80 microprocessor may be used. It will also provide a means of testing some two's complement binary calculations on paper against the solutions that the computer produces.

Whenever the microprocessor adds two numbers together one of them must be resident in the accumulator and the other may be in a register or a memory address. This means that to add two numbers together from memory, one of them must first be loaded into the accumulator. If the numbers share adjacent memory addresses it is usual to employ an indirect addressing technique in the program.

Figure 3.1 shows how two numbers from adjacent memory locations may be added together and the result stored in the next memory location.

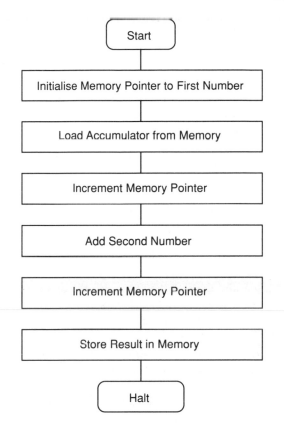

Figure 3.1 Binary arithmetic flow chart

Figure 3.1 may be translated directly into a machine code program as shown below. It assumes that the numbers to be added are in addresses 1A00 hex and 1A01 hex and that the result is to be placed into address 1A02 hex. The HL register pair is used as the address pointer.

Program

Address	Hex code		Mnemonic
			ORG 1800H
1800	21 00 1A	START:	LD HL,1A00H
1803	7E		LD A,(HL)
1804	23		INC HL
1805	00		NOP
1806	00		NOP
1807	86		ADD A,(HL)
1808	23		INC HL
1809	77		LD (HL),A
180A	76		HALT

Procedure

(a) Enter the program above into the Micro-Professor starting at address 1800 hex.

(b) Now enter the hexadecimal numbers from the following calculation into addresses 1A00 hex and 1A01 hex.

$$57 + 1B =$$

(c) Before running the program, write down the binary equivalent of the numbers above, add them together and convert the answer back to hexadecimal. This should provide the expected answer. Now run the program.

(d) Examine address 1A02 hex for the answer.
    If it is not what was expected, check the maths first and then the computer program.

(e) Complete the following calculations by converting each decimal number to binary using two's complement notation and then enter each value in hexadecimal into the appropriate addresses. Give the answer in both hexadecimal and decimal.

(i) $29 + 55$
(ii) $-10 + 93$
(iii) $-17 + (-34)$

(f) Two instructions exist in the Z80 instruction set which allow the programmer to manipulate the carry flag directly. These are:

SCF   – Set carry flag   – code 37
CCF   – Complement carry flag   – code 3F

Change the program so that the ADC A,(HL) is used in place of the ADD A,(HL).

By inserting either one or both of the instructions which manipulate the carry flag in place of the NOP's in the program, perform the following binary calculations first with the carry flag at logic 0 then with it at logic 1. Give the results in binary.

(i) $1 1 0 0 0 1 1 0 + 0 0 1 1 1 0 0 1$
(ii) $1 1 1 1 0 0 0 1 + 0 0 0 0 1 1 1 1$
(iii) $1 0 1 1 0 0 1 1 + 0 0 1 0 1 1 1 0$

Questions

3.1 How would the program have to be changed if the results of each calculation had to be sent to the output port?

3.2 How would the program have to be changed if the numbers to be added were in registers B and C and the result had to be placed in register D?

(g) Change the program again so that the ADC instruction is replaced by the subtract instruction SUB (HL). Perform the following calculations in binary.

(i) $0 1 1 1 1 0 0 0 - 0 0 1 0 1 0 1 0$
(ii) $1 0 1 1 1 1 1 0 - 1 1 1 1 1 1 1 1$
(iii) $0 0 0 0 0 0 0 1 - 1 0 0 0 0 0 0 1$

Question

3.3 Write a program starting at address 1800 hex which subtracts the number read in from port 80 hex from the value in memory address 0002 hex and leaves the result in address 1A00 hex. It then adds the value of the carry flag generated by the previous calculation and puts the result in address 1A01 hex.
    What is the lowest input number which will cause the values in 1A00 hex and 1A01 hex to be different?

Summary

The Z80 microprocessor supports a range of arithmetic operations but they are all based on simple add or subtract instructions. Either register, immediate or indirect addressing may be used to allow maximum program flexibility. The state of the carry flag may be included in calculations when the appropriate instruction is used. In addition, the carry flag state can be directly modified by two instructions so that it can be forced to either a logic 0 or logic 1 state.

PRACTICAL EXERCISE 3.2 – MULTIPLICATION AND DIVISION

(A) Multiplication

Experimental concepts One common method of multiplication is known as the '**repeated addition**' method. This works on the principle that, for example:

$$5 \times 3 = 5 + 5 + 5$$

The main disadvantage with this method is that when the numbers involved are large, the time taken to perform the calculation becomes excessive. A better method is similar to the 'long multiplication' technique often taught in schools. This is known as the '**shift and add**' method. Its maximum number of operations is equal to the number of bits in the multiplier, i.e. for an 8 bit system, only eight additions are required.

Consider the following example which illustrates how two 4-bit numbers could be multiplied.

Note: LSB means least significant bit, and MSB means most significant bit.

```
      1 1 1 0    Multiplicand
      1 0 1 1    Multiplier
      ───────
      1 1 1 0    Multiplicand × 1
    1 1 1 0      Multiplicand × 1
                 (shifted left 1 bit)
    ─────────
  1 0 1 0 1 0    Add
  0 0 0 0        Multiplicand × 0
                 (shifted left 1 bit)
  ───────────
  1 0 1 0 1 0    Add
1 1 1 0          Multiplicand × 1
                 (shifted left 1 bit)
─────────────
1 0 0 1 1 0 1 0  Add – Final product
```

A number of things are important here. At each stage, if the relevant bit of the multiplier is a 1 then the multiplicand is added to the current result, but if it is a 0 then an addition of 0 0 0 0 takes place. Clearly this addition could be eliminated.

The example shows the least significant bit of the multiplier being used first in the process. This is an arbitrary choice. In fact it is easier when the computer performs the multiplication to make it start with the most significant bit of the multiplier. This allows the most significant bit of the multip-lier to be shifted into the carry flag to test whether or not an addition should take place.

Shifting of the intermediate results must also take place so that the correct results are produced. This replaces the shifting of the multiplicand as shown in the example.

If two 8-bit numbers are to be multiplied, than a 16-bit result could occur. Therefore the registers could be assigned as shown in *Figure 3.2*. Register D contains 0 throughout the calculations.

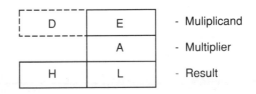

Figure 3.2 Register allocation

Program

Figure 3.3 (overleaf) can be translated into the following subroutine which may be used with any program.

Address	Hex code		Mnemonic	
			ORG 1900H	
1900	06 08	MULTIP:	LD B,8	; COUNTER
1902	16 00		LD D,0	; CLEAR D
1904	62		LD H,D	; CLEAR H
1905	6A		LD L,D	; CLEAR L
1906	29	LOOP:	ADD HL,HL	; SHIFT LEFT
1907	07		RLCA	; ROTATE A
1908	D2 0C 19		JP NC,NADD	
190B	19		ADD HL,DE	; ADDITION
190C	05	NADD:	DEC B	
190D	C2 06 19		JP NZ,LOOP	
1910	C9		RET	

Note: The ADD, HL,HL instruction shifts the contents of the HL pair one place to the left. The first time it is used it has no effect because the register pair contains all 0s. Also note the use of ADD HL,DE which is a 16-bit addition instruction in which the HL pair acts like a 16-bit accumulator.

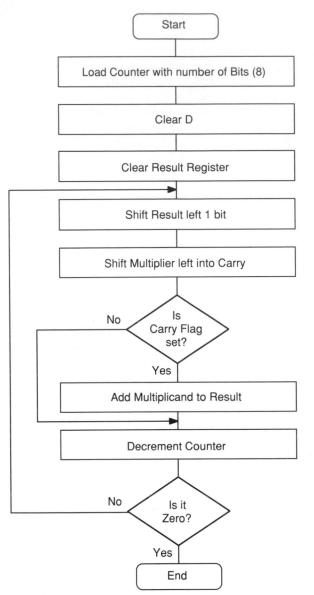

Figure 3.3   Multiplication flow chart

Procedure
The easiest way to test the multiplication program is to try it out with some simple calculations:

(a) Write a **MAIN** program that performs as follows:
   (i)   Load the multiplier into A.
   (ii) Load the multiplicand into E.
   (iii) Call the multiplication subroutine.
   (iv) Halt.

(b) Load the main program and subroutine into the Micro-Professor starting at addresses 1800H and 1900H respectively.
   At first, try some simple numbers in A and E such as 5 and 3 and use them to check the operation of the program. When the program halts, press [MONI] to regain control and then check the contents of HL.
(c) Assuming that the program works for simple numbers, try some larger ones.

Questions

3.4 Does the program work for all 8 bit numbers?

3.5 Write a **MAIN** program which performs the following calculations and leaves the result in the HL register pair

$$16 \times 7 \times 2 \times 109$$

   NB. Load E from L after the first two steps.

(B) Division

Experimental concepts Like multiplication, there are a number of methods of division which may be implemented on a computer. One is simply a repeated subtraction operation and another is similar to the shift and add method of multiplication, but is a 'shift and subtract'.

One of the best methods is the one shown here which is very similar to a 'long division' technique which would be used for decimal division. Naturally it has to be slightly modified for operation on a computer.

Consider the following problem, 17/3 where 17 is the dividend and 3 is the divisor:

$$\text{Divisor} \quad 0\,1\,1\,\overline{)1\,0\,0\,0\,1} \quad \begin{array}{l}\text{Quotient}\\ \text{Dividend}\end{array}$$

The technique involves attempting to subtract 011 from the most significant bits of the dividend. If 011 cannot be subtracted, a 0 is placed in the quotient, but if 0 1 1 can be subtracted, then a 1 is placed in the quotient and the dividend is reduced accordingly:

```
          0 0 1 0 1   – Quotient
0 1 1 ) 1 0 0 0 1
          0 1 1       – Divisor too large;   0 in quotient
           0 1 1      – Shift divisor right;  0 in quotient
            0 1 1     – Shift divisor right
          _____
          0 0 1 0 1   – Subtract;            1 in quotient
           0 1 1      – Shift divisor right;  0 in quotient
            0 1 1     – Shift divisor right
          _____
            0 1 0     – Subtract;            1 in quotient
                        remainder 0 1 0
```

The result gives 0 0 1 0 1 (5) in the quotient and 0 1 0 (2) as the remainder.

When implemented on the computer it is easier to shift the dividend and quotient left rather than shifting the divisor right (*Figure 3.4*).

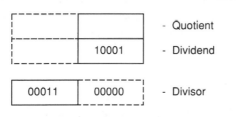

- Quotient
- Dividend

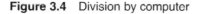

- Divisor

Figure 3.4   Division by computer

Imagine the dividend and quotient shifting left into the spaces indicated by the dotted lines.
The rule for divison is:

Shift the dividend and quotient left. If the dividend is larger than the divisor, subtract the divisor and put a 1 in the quotient. Otherwise put 0 in the quotient.

Figure 3.5 shows how this technique can be used to implement a 16-bit division routine.
The registers are used as follows:

DE holds the dividend initially
BC holds the divisor
HL is the working register for subtractions
A is the loop counter

As the calculation proceeds the quotient shifts into DE from the right.
Figure 3.5 can be translated into the program on page 36. Because of the absence of a 16 bit *com-*

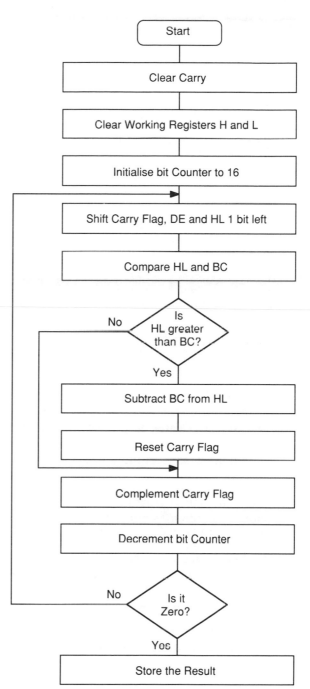

Figure 3.5   Division flow chart

pare instruction, the only way this can be performed is to perform a *subtraction*. This instruction will then set the flags and the original numbers can be restored if required by performing a subsequent *add* instruction.

Program 2

Address	Hex code		Mnemonic	
			ORG 1920H	
1920	AF	DIVIS:	XOR A	; Clear carry
1921	67		LD H,A	
1922	6F		LD L,A	; Clear HL
1923	3E 10		LD A,16	
1925	CB 13	DIVO:	RL E	; C into bit 0
1927	CB 12		RL D	; Rotate DE
1929	ED 6A		ADC HL,HL	; Rotate HL
192B	ED 42	TEST:	SBC HL,BC	
192D	D2 31 19		JP NC,DIVI	
1930	09		ADD HL,BC	; Restore HL
1931	3F	DIVI:	CCF	; Carry result
1932	3D		DEC A	
1933	C2 25 19		JP NZ,DIVO	
1936	EB		EX DE,HL	; Store result
1937	ED 6A		ADC, HL,HL	; Get last bit
1939	C9		RET	

Procedure

The division subroutine divides two 16 bit numbers and the best way to test it is to try it with some values.

(a) Write a **MAIN** program which will load a dividend into the DE register pair and the divisor into the BC register pair. It will then call the **DIVIS** subroutine and when this returns it should **halt** so that the registers may be examined.

(b) Enter the main program and the division subroutine into memory starting at addresses 1800H and 1920H respectively.

(c) Try some simple numbers in the program to start with. The main program should **halt** almost immediately it is run. Press [MONI] to regain control and then examine HL and DE.

   HL should contain the quotient and DE should contain the remainder.

(d) Try some larger numbers if the smaller ones work and make sure that there are no wrong results.

——————— Questions ———————

3.6 Can this program divide by 0? What happens?

3.7 Write down the results to the following calculations in hexadecimal:

(a) 1234H ÷ 0123H
(b) FFFFH ÷ 0019H
(c) 0H ÷ 1111H
(d) 5000H ÷ 5001H

3.8 Write a **main** program which performs the following calculation:

$$(9000H \div 820H) \div 13H$$

Summary

Multiplication and division are two mathematical functions which are not normally supported by 8 bit microprocessors. Simple programs have to be written that will perform these calculations using only these functions that are available i.e. addition, subtraction and shifting of data.

Once written as subroutines, multiplication and division programs may be used as the basis of routines to calculate many other more complex mathematical functions. In this way, many high-level computer languages such as BASIC and PASCAL are capable of performing highly complex mathematic calculations.

PRACTICAL EXERCISE 3.3 – MULTI-BYTE ARITHMETIC

Experimental Concepts

When numbers are represented in binary in a computer the number of bits used places a limit on the range of values which can be handled. For example, if 8 bits are used, the range of values is 0 to 255. Similarly, with 16 bits the range of values is 0 to 65535. If higher numbers need to be accommodated then the number of bits must be increased accordingly.

Consequently many calculations have to be performed by computers on numbers which may occupy more than one byte of memory and so the need arises for some multi-byte arithmetic capability.

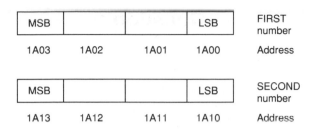

Figure 3.6 Multi-byte numbers

The Z80 does provide some 16 bit arithmetic instructions but the range is very limited. It is possible, however, to write short programs which can handle any number of bytes of data with relative ease. Since these large numbers cannot be stored efficiently within the microprocessor registers it is necessary to allocate a number of memory locations for each one. This is shown in *Figure 3.6*.

Two four byte numbers are shown occupying addresses 1A00–1A03 hex and 1A10–1A13 hex respectively. Each one is capable of storing numbers up to 4295 million in binary.

When the addition or subtraction operation takes place, it is carried out starting with the least significant byte of each number and progressing with each pair of bytes until the most significant pair has been operated upon. The only real problem is that when a carry is generated by the arithmetic process it must be taken into consideration when the next pair of bytes is being handled.

Figure 3.7 shows how two multi-byte numbers may be added together. The result is put back in the same location as the FIRST number.

Program

Figure 3.7 can be translated into the program below if the CPU registers are used as follows:

B — byte counter.
DE — pointer to FIRST number.
HL — pointer to SECOND number.

Assume that four bytes are to be added, the FIRST number starts at location 1A00 hex and the SECOND at 1A10 hex.

Note that the XOR A instruction is one of the logical instructions that is used here to clear the accumulator and carry flags.

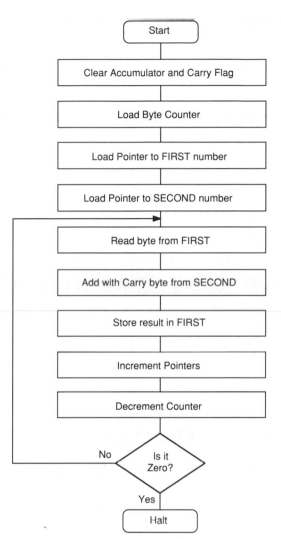

Figure 3.7 Multi-byte arithmetic flow chart

	Mnemonic	*Comment*
	ORG 1800H	
	XOR A	; Clear A and carry flag
	LD B,04	; 4 bytes to be added
	LD DE,1A00H	; DE points to FIRST
	LD HL,1A10H	; HL points to SECOND
LOOP:	LD A,(DE)	; Get byte of FIRST
	ADC A,(HL)	; Add byte of SECOND
	LD (DE),A	; Store result in FIRST
	INC DE	; Move pointers
	INC HL	;
	DEC B	; Decrement counter
	JP NZ,LOOP	; Repeat until done
	HALT	

Procedure

(a) Translate the program into a machine code program starting at address 1800 hex and load or download it into the Micro-Professor.

(b) Load the two four-byte numbers to be added into addresses 1A00–1A03 hex and 1A10–1A13 hex.

   Initially try the hexadecimal numbers

   17B38ED5
   and 25F7216A

(c) Work out the result to expect by adding the numbers together on paper.

(d) Now run the program. It should halt almost immediately, so press [RESET], then check the answer by looking at addresses 1A00–1A03 hex.

If it is not what you predicted then check the answers at the back of this book to see whether the prediction or the computer is correct.

Questions

3.9 What are the results of the following hexadecimal calculations?

   (a) 12345678 + A9876543
   (b) 4FFFFFE + 5
   (c) 3029D + CFE3

3.10 What happens if the result of the calculation is larger than the maximum number that can be stored in the four bytes of memory?

3.11 What is the maximum number of bytes that could be added by this technique?

(e) Now replace the ADC A,(HL) instruction with an SBC A,(HL) instruction. This changes the program so that it performs a multi-byte subtraction of FIRST − SECOND.

Question

3.12 What are the results of the following hexadecimal calculations?

   (a) 5AA508B2 − 368B9CD4
   (b) 1834E75 − 1834E76
   (c) C3200 − 302

The concept of performing multi-byte arithmetic can be extended to more than two numbers, as can be seen in the following question.

Question

3.13 Write a program that is capable of adding six 4-byte numbers located in memory at the following locations each with the least significant byte first.

   Number 1 – 1A00–1A03 hex
   Number 2 – 1A10–1A13 hex
   Number 3 – 1A20–1A23 hex
   Number 4 – 1A30–1A33 hex
   Number 5 – 1A40–1A43 hex
   Number 6 – 1A50–1A53 hex

The result should be placed in the same location as number 1. Start the program at address 1800 hex.

   Hint: Use the C register as a 'number' counter.

Summary

Multi-byte arithmetic is required in many microprocessor applications that uses numbers greater than 255. Since the Z80 has no simple instructions to deal with more than 16 bits at a time, a short program must be written to handle larger numbers. The same program can be made to process numbers of up to 256 bytes each so the number range can be very extensive.

PRACTICAL EXERCISE 3.4 – BCD ARITHMETIC

Experimental Concepts

The registers and memory addresses in an 8-bit microprocessor system hold data in binary form which generally represents numbers between 0 and 255. Other representations are possible, one of which is known as **binary coded decimal**, in which each 8-bit store is said to contain two 4-bit numbers. Each 4-bit number is restricted to those between 0 and 9, so the 8 bits are restricted to those between 00 and 99. Bits 0–3 hold the least significant digit and bits 4–7 hold the most significant digit. This is known as '**packed BCD**' format.

Unfortunately, if numbers in packed BCD format are added or subtracted, the microprocessor ALU regards them as pure binary numbers and treats them accordingly. The result is a correct binary calculation but an incorrect decimal calculation. Therefore, by inserting the decimal adjust accumulator (DAA) instruction into a program soon after the arithmetic operation the microprocessor can correct the binary result to a true decimal one.

This exercise is in two parts. The first examines the operation of the DAA instruction with various inputs and arithmetic processes. The results can best be seen by sending them to the output port lights. In the second part a multi-byte addition program similar to that discussed in the previous exercise is used to perform simple decimal calculations.

The program below may be used to investigate the operation of the DAA instruction. It simply adds two numbers together then corrects the result for decimal numbers.

Program 1

Address	Hex code	Mnemonic
		ORG 1800H
1800	AF	XOR A
1801	3E n	LD A,n
1803	06 m	LD B,m
1805	80	ADD A,B
1806	D3 81	OUT (81H),A
1808	27	DAA
1809	D3 81	OUT (81H),A
180B	76	HALT

Procedure

(A) Operation of the DAA instruction

(a) Program 1 is a very simple way of examining the DAA instruction. It adds together the two numbers held in registers A and B and then outputs the result to port 81H. It then corrects the calculation for decimal values and outputs the second result.

If the program runs at full speed, the first output instruction happens so quickly that it is impossible to see the uncorrected value. Therefore the program is only intended to be run by single stepping through it. For this exercise, use the STEP key on the Micro-Professor keyboard to run the program.

The first instruction is a useful way of clearing the accumulator and the carry flag. Enter the program into the Micro-Professor in the normal way. Initially, put in the values below for n and m.

$$n = 04$$
$$m = 06$$

(b) Go to the start of the program by pressing

[ADDR] [1] [8] [0] [0]

and then press [STEP] five times. The lights on the applications board should show the result of the calculation, which is 0A hex.

(c) Now press [STEP] twice more, and the lights should change to the corrected decimal value which is 10.

Remember that bits 0–3 represent the least significant digit, and bits 4–7 represent the most significant digit.

(d) Replace the numbers n and m in the program with those indicated below and complete the table. The correction is the hex number which has to be added to the hex sum to obtain the decimal sum.

n	m	Hex sum	Decimal sum	Correction
44	72			
80	19			
88	08			
67	35			

(e) Change the ADD A,B instruction to a SUB B instruction (op-code 90 hex) and complete the table below

n	m	Hex difference	Decimal difference	Correction
76	28			
82	41			
11	22			
33	42			

It may be necessary to perform a simple calculation on paper to confirm the result.

Now try the question below.

Question

3.14 Which of the calculations above will cause the **carry** flag to be **set** after the operation of the DAA instruction?

(B) Multi-byte decimal addition
Multi-byte calculations can be performed in decimal in almost the same way as in binary with the inclusion of a DAA instruction in the program loop. The multi-byte binary addition program examined in Practical Exercise 3.3 is shown below suitably modified for decimal calculations. Note the use of the **add with carry**, ADC A,(HL) instruction in the program so that the carry from the previous addition is added into each new pair of digits.

Program 2

Address	Hex code		Mnemonic	Comment
			ORG 1800H	
1800	AF		XOR A	; Clear A and carry
1801	06 03		LD B,03	; Number of bytes
1803	11 20 18		LD DE,1820H	; First number
1806	21 30 18		LD HL,1830H	; Second number
1809	1A	LOOP:	LD A,(DE)	
180A	8E		ADC A,(HL)	
180B	27		DAA	
180C	12		LD (DE),A	
180D	13		INC DE	

Address	Hex code	Mnemonic	Comment
180E	23	INC HL	
180F	05	DEC B	
1810	C2 09 18	JP NZ,LOOP	
1813	76	HALT	
1820	96	LSB	First number
1821	84		
1822	23	MSB	
1830	35	LSB	Second number
1831	21		
1832	14	MSB	

(a) Enter the program and data shown above into the Micro-Professor.
(b) Work out the result to expect on paper, then look at the program and decide where the result will be found in the computer.
(c) Run the program. Now locate the result in the computer memory and check to see if it is what was expected.
(d) Run the program again. What is the new result?
(e) Use your program to add the following numbers:

(i) 7496 + 1085   (iii) 3009 + 82
(ii) 2739128 + 145917

Questions

3.15 Does **program 2** work equally well for decimal subtraction if the ADC instruction is replaced with an SBC instruction?

Summary

The Z80 microprocessor instruction set includes a very powerful decimal adjust accumulator instruction. With this it is possible to perform calculations on numbers which are stored in binary coded decimal format within memory. Both addition and subtraction operations are automatically corrected to the true decimal values. Multi-byte operations are also possible in a simple loop program as long as the carry bit generated by each addition or subtraction is included in the next arithmetic process.

Control systems

EQUIPMENT REQUIRED

To complete all of the practical exercises in this chapter you will need:

(a) *Micro-Professor MPF-1B.*
(b) *Applications board MAB.*
(c) *The POLSU replacement EPROM set.*

It is also advisable to use a cross-assembler on an IBM PC or compatible computer to write the programs in assembly language. This will also require:

(d) *PC with cross-assembler, editor, linker and connecting lead.*
(f) *(Optional) practical exercises on disk.*

PREREQUISITES

Before studying this chapter you should have:

(a) *An understanding of the complete Z80 instruction set in hexadecimal and mnemonic format.*

(b) *An awareness of the need for programmable user ports in a microcomputer system.*
(c) *An awareness of the concept of a feedback control system.*
(d) *An appreciation of the use of subroutines.*

These may be obtained by studying the first three and the fifth chapter of the companion book in the series, *Microelectronics NII*.

4.1 INTRODUCTION

Most microprocessors are employed to control something. It may be the carriage of a printer or a car engine, a washing machine or a steel mill: in one way or another microprocessors are gradually taking over the majority of the control functions in the industrialised world. They perform their operations by sending binary codes to interface circuits which turn various actuators on and off to control the devices to which they are connected.

When a process has to be carefully regulated, circuits are normally arranged so that the system sends information back to the microprocessor to inform it of its current position or operation. This feedback can then be used by the software in the microcomputer to modify any control signals previously sent.

In this chapter, some simple control systems are introduced which are both 'open loop' (i.e. without feedback) and 'closed loop' (i.e. with feedback).

The motor on the applications board provides a very visual indication of the operation of a control system. The exercises involve sending signals

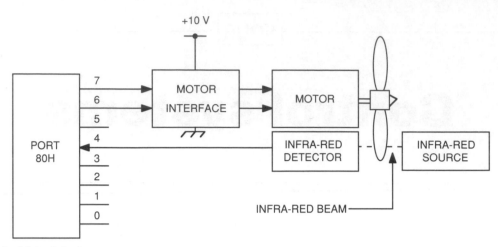

Figure 4.1 Motor interface

from the microprocessor to the motor to make it perform various functions including turning for a fixed time or turning a set number of revolutions.

Feedback signals are provided to the computer because the propeller on the motor shaft cuts an infra-red beam three times in every revolution.

An alternative feedback system using analogue control is examined in the final exercise. Here the temperature of a sensor is detected and converted from an analogue voltage to its digital equivalent so that a heater can be switched on or off to maintain the desired temperature. This is the basis of a simple thermostat.

PRACTICAL EXERCISE 4.1 – MOTOR CONTROL

Experimental Concepts

The object of this exercise is to use the microprocessor to control a d.c. motor. *Figure 4.1* shows how the motor and its interface electronics are connected to the Micro-Professor.

Notice that all the connections are made with port 80H. This port is also connected to the 8 input switches which allows the motor to be controlled manually when required. However, when the motor is under computer control the input switches can generally be ignored. It is best to leave them in the logic 1 (OFF) position.

Port 80H is a **programmable** port. This means that each bit can be selected to be either an **input** or an **output**. Whenever the RESET key is pressed,

all the port 80H bits revert to being **inputs** if the POLSU EPROM set is installed. In this experiment however, bits 6 and 7 must be **outputs**. This is accomplished by sending certain codes to the port in an **initialisation** program. The program required is given without a great deal of explanation since the subject of programming input/output ports is covered in detail later.

Once bits 6 and 7 have been established as outputs, the codes shown in *Figure 4.2* can be sent to the port to control the motor. Notice that only bits 6 and 7 are actually output although a complete byte has to be sent to the port. Bits 0–5 could be anything, although it is usual to make them all 0s.

The program opposite shows how the port initialisation instructions and a simple motor control instruction may be combined.

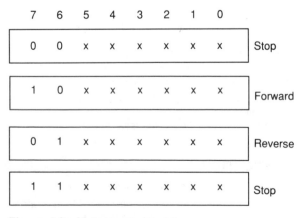

Figure 4.2 Motor control codes

Program

Address	Hex code		Mnemonic
			ORG 1800H
1800	3E FF	INIT:	LD A,0FFH
1802	D3 82		OUT (82H),A
1804	3E 3F		LD A,3FH
1806	D3 82		OUT (82H),A
1808	3E 80	FWD:	LD A,80H
180A	D3 80		OUT (80H),A
180C	76		HALT

When executed, the first four instructions set up the port for motor control, then the last three instructions make the motor turn in the forward direction.

Procedure

(A) Simple motor control

(a) Make sure that all the port 80H switches are up and that all the other switches on the applications board are OFF except:

MOTOR – ON

(b) Enter the program into the Micro-Professor and execute it. The motor should start to turn at high speed.

(c) To stop the motor temporarily at any time simply switch the MOTOR control switch OFF.

Press [MONI] to return control to the monitor program. *Do not press* [RESET] or the motor will stop.

(d) Replace the code in the program that makes the motor go forward with one that will stop it. Make sure that the MOTOR control switch is ON. Now execute the new program. The motor should stop.

(e) Change the motor control code to the one that will make the motor reverse and execute the program again. The motor should go in reverse.

(f) Press [RESET] to stop it. This also makes all the bits of port 80H revert to being inputs.

(g) The significance of the port initialisation program can be seen by trying to control the motor without first initialising the port. To do this, execute the program starting at address 1808 hex. Does anything happen?

Compare this with the results of executing the whole program from address 1800 hex.

Remember, the port initialisation part of the program need only be run once at the beginning of any other programs which control the motor.

Question

4.1 Look again at the program. Which **logical** instruction could have been used at address 1808 hex to make the motor go forward by forcing bit 7 to a logic 1?

Which **logical** instruction would make it go in reverse?

(B) Creating a time delay

It is quite often necessary to make the microprocessor carry out certain operations for a fixed length of time. This is achieved by sending the starting command for the operation and then making the microprocessor delay in a counting loop.

Consider the program below:

```
        LD B,00
LOOP: DEC B
        JP NZ,LOOP
```

This represents a simple counting loop which will cause a delay of about 2 ms. The exact time delay can be calculated by multiplying the number of loops (256 in this case) by the time per loop. The time per loop depends upon the number of T-states required by each instruction. In the example above:

LD B,00	– 7 T-states
DEC B	– 4 T-states
JP NZ,LOOP	– 10 T-states

Thus the total program requires

$$7 + 256 (4 + 10) = 3591 \text{ T-states}$$

The Micro-Professor clock rate is 1.79 MHz so each T-state takes 560 ns.

Total time for the program is

$$3591 \times 560 \text{ ns} = 2.01 \text{ ms}$$

A longer time delay can be created by counting down using a register pair as the counter instead of a single register. The only drawback with this method is that the decrement instruction for a register pair **does not** set the flags. In particular another means must be found to discover when the count has reached zero. This technique is shown in the program below.

```
            LD BC,0000
   LOOP:    DEC BC
            LD A,B
            OR C
            JP NZ,LOOP
```

The two instructions

```
            LD A,B
            OR C
```

are designed to perform the logical OR function between registers B and C, the counter registers. If a logic 1 exists anywhere in the register pair then the zero flag will not be set. However when the count reaches 0000, the zero flag will be set.

The instructions in the previous program take the following number of T-states:

```
      LD BC,0000   –  10
      DEC BC       –   6
      LD A,B       –   4
      OR C         –   4
      JP NZ,LOOP   –  10
```

Calculate the length of the delay created.

Now test the time delay by proceeding as follows:

(a) Write a program that:
    (i)   initialises the port for motor operation.
    (ii) makes the motor go forward.
    (iii) delays for the time given by the previous program.
    (iv) makes the motor stop.

(b) Enter the program and execute it. Does it run for as long as expected?

(c) An EXCLUSIVE OR instruction can be used to invert any chosen bits in a data byte.
    What instruction would invert bits 6 and 7 but leave the other bits unaffected?

(d) Modify the program written in (a) above so that it:
    (i)   Initialises the port for motor operation.
    (ii) Makes the motor go forward.
    (iii) Delays for a time.
    (iv) Reverses the direction of the motor.
    (v) Jumps to (iii).

Note that the motor control code will need to be stored in a register other than those used in the delay program.

——— Questions ———

4.2 Write a program that turns the motor for about 9 seconds in each direction before reversing.

4.3 Write a program in which the motor is controlled by the state of bits 0 and 1 of port 80H.
Note: To do this the data from port 80H will have to be input and then rotated right twice before it is output to control the motor. Choose the same motor control codes as given previously.

Summary

When bits 6 and 7 of port 80H are arranged to be outputs by running an initialisation program, they may be used to control the small motor on the applications board. When these bits have different logic states the motor turns but when they are the same it stops.

Time delay programs can be created by making the computer count down from a large value. Unfortunately, the instruction that decrements a register pair does not set the flags, so an additional logical instruction must be inserted to achieve this. An exclusive OR instruction may be used to reverse the direction of the motor.

PRACTICAL EXERCISE 4.2 – BIT MASKING AND TESTING

Experimental Concepts

Selecting individual bits of data from a byte is a frequently required operation.

When a Z80 microprocessor system inputs data from a port it always receives 8 bits. Very often only a few of those bits are of interest and so the system must have a way of ignoring unwanted information. This process is known as '**masking**' the unwanted bits and is easily accomplished with the LOGIC instructions within the microprocessor instruction set. In the same way it is possible to use the logic instructions to test the condition of one or more bits of an input port. This makes it possible to take a certain course of action when the input conditions are correct.

Bit masking is normally performed with the AND instruction since this can be made to force unwanted bits to a logic 0 state. This can also select one or more bits from a byte by eliminating all the others. When the required bit or bits have been selected from a byte of data any testing can be carried out with a conditional jump instruction, generally jump on zero or jump on non-zero. Occasionally a logic OR instruction may be used to force unwanted bits to a logic 1 state.

Two major possibilities follow from the idea of bit masking and testing. The first is that it is possible to wait for certain input conditions to be present before performing some function. The second is that the computer can continuously check an input until something changes. Both of these operations will be considered in this exercise. In each case the routines from the previous exercise which control the motor on the applications board will be used as a typical example of an output action.

Consider *Figure 4.3* which is the basis of a program to run the motor only if bit 2 of port 80H is a logic 1. When all the bits apart from bit 2 are forced to logic 0 by the masking process and if the result is zero (the zero flag is set) then bit 2 must have been a logic 0 when it was input. However, if the zero flag is not set because all bits are not logic 0 then bit 2 must have been a logic 1 when it was input.

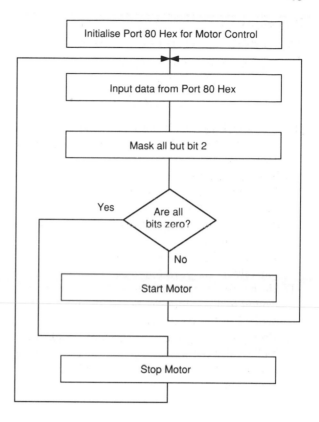

Figure 4.3   Bit masking flow chart

Figure 4.3 can be translated into Program 1 below.

Program 1

	Mnemonic	Comment
	ORG 1800H	
INIT:	LD A,0FFH	; Motor initialisation
	OUT (82H),A	
	LD A,3FH	
	OUT (82H),A	
START:	IN A,(82H)	
	AND 04H	; Mask all but bit 2
	JP Z,STOP	
FWD:	LD A,00H	; Bit 2 must have been 1
	OUT (80H),A	; Start motor
	JP START	
STOP:	LD A,00	; Bit 2 must have been 0
	OUT (80H),A	; Stop motor
	JP START	

Procedure

(A) Testing for certain input conditions

(a) On the applications board switch all the switches OFF apart from:

MOTOR – ON

(b) Convert Program 1 into machine code and enter it into the Micro-Professor at address 1800 hex.

(c) Run the program.
If the bit 2 switch of port 80H is at a logic 1 the motor should run.
If not, try switching the bit 2 switch and see what happens.
Try the other switches and see if they have any effect.

(d) Modify the program so that the motor is controlled in the same way but by the bit 3 switch instead of the bit 2 switch.

(e) Modify the program again so that a logic 0 on bit 3 makes the motor run and a logic 1 makes it stop. Only one byte will need to be changed to do this.

—————— **Questions** ——————

4.4 How can the program be modified so that a logic 1 on bit 2 OR on bit 3 will cause the motor to run? It will only stop if they are both logic 0.

4.5 How can the program be modified so that a logic 1 on bit 2 AND on bit 3 will cause the motor to run? It will stop if either or both are at logic 0. (An extra instruction is required.)

(B) Waiting for an input to change

If data from an input port is continuously compared with the previous data from that port it should be possible to determine whether the data has changed. This is the idea embodied in *Figure 4.4*. It is designed to cause the motor to either start or stop whenever an input bit changes.

If the B register is used for the original or old

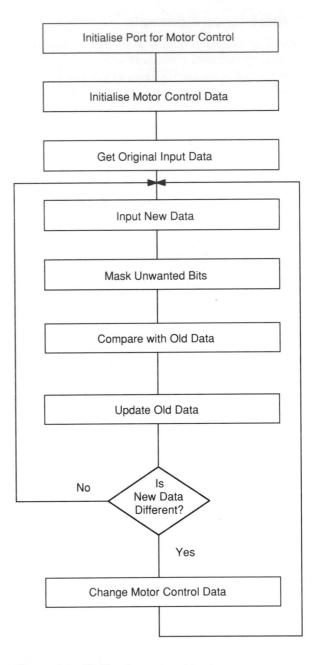

Figure 4.4 Waiting for an input to change

data and the C register is used for the motor control data, the following program can be produced. The motor should stop or start whenever any one of bits 0–3 of port 80H is changed.

Program 2

	Mnemonic	Comment
	ORG 1800H	
INIT:	LD A,0FFH	; Initialise port
	OUT (82H),A	
	LD A,3FH	
	OUT (82H),A	
	LD C,80H	; Motor FWD code
GETDAT:	IN A,(80H)	
	AND 0FH	; Mask unwanted bits
	LD B,A	; Store in B
LOOP:	IN A,(80H)	; Get new data
	AND 0FH	; Mask unwanted bits
	CP B	; Compare old data
	LD B,A	; Update old data
	JP NZ,MOTA	; Jump if different
	JP LOOP	
MOTA:	LD A,C	; Get motor code
	XOR 80H	; Change bit 7
	LD C,A	; Store code
	OUT (80H),A	
	JP LOOP	

Now proceed as follows:

(a) Translate program 2 into machine code and enter it into the Micro-Professor starting at address 1800H.

(b) Make sure the motor control switch is still ON then run the program. Now change one of the bits 0–3 on port 80H.

Change the other switches on the port. If it does not behave as it should, go back and check the program.

Now answer the following questions.

Questions

4.6 In program 2, is the instruction LD C,80H really necessary? What happens without it?

4.7 In Program 2, are the instructions below necessary?

GETDAT: IN A,(80)
           AND 0FH
           LD B,A

What happens if they are removed?

4.8 Why is the B register updated in program 2 **before** the conditional jump instruction JP NZ,MOTA?

The following two questions are based on the previous programs. If time permits, try the programs on your computer just to check their operation.

Questions

4.9 A program is required which outputs the data F0 hex to port 81H if the following data is present on port 80H.

Bit 3 – logic 0
Bit 4 – logic 1
Bit 5 – logic 0
Bit 6 – logic 1

All other bits can be either 1 or 0.
If the data is incorrect the system outputs 0F hex. Write a suitable program starting at address 1800H.

4.10 Write a program that increments the number displayed on the lights of port 81H whenever an input bit from port 80H changes state. Start the program at address 1800H.

Summary

Logic instructions can be used in many ways but one of their main applications is to force bits into known states. The logic **AND** instruction is particularly useful in forcing bits to a logic 0 state.

This is known as 'bit masking'. The same instruction can isolate certain bits from a byte of data so that individual bits may be tested and appropriate action taken.

If input data from a port is continuously compared with the previous data value input, it is possible to monitor any change of state of the inputs. Thus programs can easily be created which wait for an input condition to change before any action is taken.

PRACTICAL EXERCISE 4.3 – EVENT COUNTING

Experimental Concepts

The concept of making the computer wait for an event to happen has many applications. In fact, many computers spend the majority of their time waiting for human inputs to occur.

In this exercise the idea of waiting for the propeller on the motor shaft to cut the infra-red beam will be used as a means of counting the number of revolutions of the motor.

When the three-bladed propeller on the applications board cuts the infra-red beam, a logic 0 is generated and when the beam is unbroken, a logic 1 is produced. Therefore, as the propeller rotates a waveform like that shown in *Figure 4.5* is generated where three negative-going pulses represent one revolution.

Suppose a program is required which detects the passage of each propeller blade. This means that it has to wait for a logic 0 to occur, then wait for a logic 1 which would indicate a single propeller blade. The detector input is on BIT 4 of port 80H. A suitable flow chart is shown in *Figure 4.6*.

The first loop waits for a logic 0 and when this occurs the program proceeds with the second loop which waits for the logic 1.

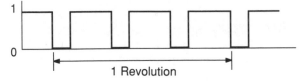

Figure 4.5 Output from infra-red detector

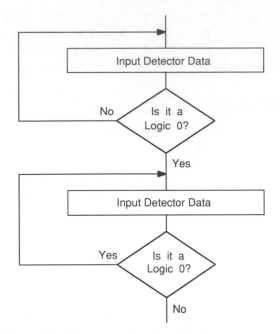

Figure 4.6 Event counting flow chart

If this program is used as part of a routine that counts three such events, then this represents a revolution counter.

Program 1

Figure 4.6 has been used as the basis of the revolution counting subroutine. When used in a main program, the subroutine CALL can be treated as 'wait for one revolution' since control does not return to the main program until the infra-red beam has been cut three times. Register B holds the value for the number of times the beam has to be cut.

	Mnemonic	*Comment*
REV:	LD B,03	; 3 BLADES
LOOP1:	IN A,(80H)	
	AND 10H	
	JP NZ,LOOP1	; WAIT FOR 0
LOOP0:	IN A,(80H)	
	AND 10H	
	JP Z,LOOP0	; WAIT FOR 1
	DEC B	
	JP NZ,LOOP1	
	RET	

Procedure

Writing a program can be made very much easier if it can be divided into different parts. Conveniently, these can be subroutines. Remember, a subroutine must finish with a **return** instruction, and because of this it **cannot** be run as a program on its own. Subroutines must be **called** from another program. If a subroutine is run on its own, the Micro-Professor will execute it once, then the display will show SYS-SP.

Proceed as follows:

(a) Write a subroutine, which will initialise the PIO for motor control. This will be identical with the program given previously. Locate this subroutine at address 1800H.

(b) Write a subroutine which will make the motor go forward. Locate this at address 1810H.

(c) Write a subroutine which will make the motor go backwards. Locate this at address 1820H.

(d) Write a subroutine which will make the motor stop. Locate this at address 1830H.

(e) Write a subroutine which will cause a delay of about 1 second. Locate this at address 1840H.

(f) Write a subroutine given in program 1 which will count one revolution of the propeller. Locate this at address 1850H.

These subroutines can be regarded as the building blocks for a number of programs. They can be called as many times as required in the main program.

In assembly language, multiple origins can be accommodated simply by including a separate ORG directive at the beginning of each subroutine.

For example, consider the following program which turns the motor for one revolution in each direction alternately.

Program 2

This program assumes that the subroutines listed in steps 1 to 6 have already been written.

	Mnemonic	Comment
	ORG 1870H	
MAIN:	CALL 1800H	; PORT INITIALISATION
LOOP:	CALL 1810H	; FORWARD
	CALL 1850H	; 1 REVOLUTION
	CALL 1830H	; STOP
	CALL 1840H	; DELAY
	CALL 1820H	; REVERSE
	CALL 1850H	; 1 REVOLUTION
	CALL 1830H	; STOP
	CALL 1840H	; DELAY
	JP LOOP	; REPEAT

(a) Convert program 2 into machine code and enter it into the Micro-Professor starting at address 1870H. If assembly language is being used, the addresses can be replaced by labels.

(b) Make sure that bits 7 and 6 of the port 80 switch are in the logic 1 position (up) and that the motor control switch is ON (down).

Now **run** the program – but remember that it starts at address 1870H.

If the display shows SYS-SP, check that all the subroutines end with a RETURN instruction.

Don't switch off the computer. All the subroutines are required again for the following exercises.

Exercises

Program 2 used one CALL instruction after another. Generally they would be interspersed with other lines of program. This is what is required in the following questions.

———— Questions ————

4.11 Modify the previous program so that the motor turns 5 revolutions in each direction before reversing. Either overwrite the previous program or start it at some convenient address further on in memory. Do not just **call** the revolution counting subroutine five times.

4.12 It is often necessary to take different actions as a result of differing input conditions. Modify the program from Question 4.11 so that the motor turns in each direction by the number of revolutions set on bits 0–3 of the port 80H switch. This will mean that a binary number between 0 and 15 can be input. Be careful that to arrange the program so that an entry of 0 causes no motor movement, rather than 256 revolutions and that all bits are masked apart from those required. Start it at address 1900H.

Summary

Event counting is a widely used technique and has very general application in control systems. When events can be signalled to the microprocessor in the form of a digital pulse, care must be taken that only one event is signalled for each pulse. This means that the software must wait for one leading and one trailing edge of each pulse.

The use of subroutines greatly simplifies programs and many program variations are made possible simply by using the subroutines in different ways. In addition, certain types of subroutine lend themselves to parameter passing (see *Microelectronics NII*, Chapter 5) and this also increases their flexibility.

PRACTICAL EXERCISE 4.4 – THERMOSTATIC CONTROL SYSTEM

Experimental Concepts

Since it is possible to CALL a subroutine from any other program, this means that subroutines that already exist in the system ROM may be used by other programs in order to save rewriting similar routines. Clearly it is necessary to understand fully what the system subroutines do, what registers they affect, what the input and output requirements are, etc. However, a little information on each one will allow them to be used in many different ways.

This exercise introduces a number of system routines which will allow programs to be written

to simulate a simple thermostatic control system with a built-in alarm.

The heater on the applications board is controlled by bit 5 of port 80H; when this is a logic 1 the heater is ON and when it is a logic 0 the heater is OFF. Before it will operate, however, the heater control switch must also be turned ON, and port 80H must be initialised for heater control. This initialisation procedure is very similar to that employed in the motor drive exercises and a subroutine has been included which can be called to save having to write it.

Monitor Subroutines

In the following exercise the monitor subroutines listed below may be used in much the same way as those written as part of the normal program. Be careful to note the input and output conditions of each one as well as the main function. It is very important with all monitor routines that the programmer knows exactly what they will do, and which registers they affect. If the subroutine destroys the contents of any register which is being used in the program to store a special value, simply **push** the contents of the register pair onto the stack before calling the subroutine then **pop** the contents when the subroutine returns.

(a) Name: MHINI
   Address: 209C hex
   Function: Initialise Port 80H for heater and motor control
   Input: None
   Output: None
   Registers destroyed: None

(b) Name: ANALOG
   Address: 2001 hex
   Function: Convert the analogue input into an 8-bit digital value
   Input: None
   Output: Register A contains the digital equivalent of the analogue input between 00 and FF hex.
   Registers destroyed: AF

(c) Name: DISPA
Address: 2420 Hex
Function: Display the contents of register A as a pair of hexadecimal digits in the two right-hand displays.
Input: A contains number to be displayed.
Output: None (apart from display)
Registers destroyed: None
Note: This subroutine takes about 3 ms to execute.

(d) Name: ALARM
Address: 245E hex
Function: Beep speaker and flash two right-hand displays for about 1 second.
Input: A contains number which is displayed
Output: None
Registers destroyed: AF
Note: This subroutine takes about 1 second to execute.

A simple thermostat operates so that it switches a heater ON when the temperature measured is below a certain threshold and switches it OFF if the temperature is above the threshold. Many thermostats incorporate some hysteresis, which means that they switch on and off at different temperatures, but in the simple case this can be ignored. An over-temperature alarm may also be ignored for the moment. The simple flow chart is shown in *Figure 4.7*.

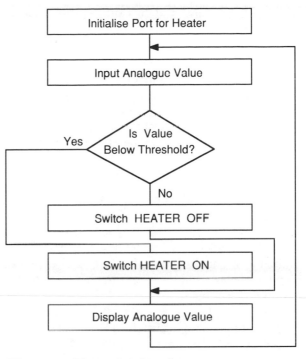

Figure 4.7 Thermostat flow chart

	Mnemonic	Comment
	ORG 1800H	
THERM:	CALL MHINI	
LOOP:	CALL ANALOG	
	CP 80H	; 80 hex is threshold value
	JP C,HON	
HOFF:	PUSH AF	
	LD A,00	
	OUT (80H),A	
	POP AF	
	JP DIS	
HON:	PUSH AF	
	LD A,20H	
	OUT (80H),A	
	POP AF	
DIS:	CALL DISPA	; Display analogue value
	JP LOOP	

Program

The program opposite is based on *Figure 4.7*. Note that a number of extra instructions have been added. A threshold value of 80 hex has been used in line 4.

Procedure

(A) Simple Thermostat program

(a) First translate the instructions of program 1 into machine code starting at address 1800 hex, and load or download it into the Micro-Professor in the usual way.

(b) Make sure the switches on the applications board are in the following positions.

MOTOR	control	OFF `
HEATER	control	ON
ADC	control	ON
BAR GRAPH	control	OFF
ANALOGUE	input to position 2, TEMPERATURE	
PORT 80H	switches ALL at Logic 1 (UP)	

(c) Now execute the program.

The display should show a two digit hexadecimal number and the heater indicator light should come ON if the number displayed is below 80 hex.

If the number is below 80 hex, as it should be initially, it should gradually start to rise as the heater begins to increase the temperature of the temperature sensor.

When the display reaches 80 hex, the heater indicator light should switch OFF. This shows that the program is working correctly.

The indicated temperature may rise slightly above 80 hex because of the thermal capacity of the heater and temperature sensor, but soon it will start to fall as it cools in the air. When the indicated value is less than 80 hex the heater should come on again, as shown by its indicator light and the temperature value will start to rise again. Thus the temperature cycles about the value set as the threshold. This temperature cycling is a characteristic of all systems that employ a thermostat for simple ON/OFF control. Much finer control of temperature can be achieved with more complex control systems but these are beyond the scope of the present exercise.

Remember that although the value returned by the temperature sensor is related to the actual temperature of the heater, the true temperature in degrees Celsius can only be obtained and displayed if the program includes the correct conversion routines.

(d) The program may be started with a **hot** sensor rather than from **cold** if the following procedure is adopted.

Press [RESET] to stop the program. If the bit 5 switch of port 80H is still in the logic 1 position the heater indicator should still be ON. If not, make sure that the heater control switch is ON and bit 5 of port 80 is a logic 1.

Now, leave the heater on for about 4 minutes. This gives it plenty of time to heat up.

Execute the program again. This time the temperature value indicated should be high and the heater indicator light should be off. The temperature gradually falls until it reaches a value of 80 hex when the heater will turn on again.

This technique of starting **hot** can be used to check the operation of the alarm in the next part of the exercise.

Now answer the following questions.

——— Questions ———

4.13 What are the highest and lowest temperature values recorded after the system has stabilised? This will give some idea of the accuracy of this simple control system.

4.14 What happens if the threshold value is changed? Try higher and lower values in the program. Over what range of threshold values does the system perform satisfactorily?

Briefly explain the reasons for the limits of satisfactory performance.

(B) Thermostat with a built-in alarm

When a microcomputer controls a process it can often be made to check for operation within pre-set limits as well as maintain the normal control action. Thus, if the program for the simple thermostat is extended it should be possible to include an over-temperature alarm which would indicate abnormal operating conditions. Under normal circumstances, of course, the computer will try to maintain the correct conditions so that the alarm would never be triggered. The alarm part of the program can only be tested if the computer is tricked in some way.

The flow chart for the thermostat with a built-in alarm is shown in *Figure 4.8*. This time there are three possible actions and two threshold values, one for normal control action and the other for the alarm. Proceed as follows, using a value of 80 hex for the control threshold, and 90 hex as the alarm threshold:

(a) Use *Figure 4.8* to write a machine code program, starting at address 1800 hex which will operate as described previously.
(b) Check the program thoroughly, particularly the jump addresses and the CALLS and then enter it into the Micro-Professor.
(c) Run the program and check that it controls the temperature of the sensor in the same way as the previous program.
(d) Now reset the computer and allow the heater to heat the temperature sensor for about four minutes.
(e) Run the program again. This time, if the sensor is sufficiently hot, the alarm should sound and the display flash. If not, check that the temperature value indicated is above 90 hex, then re-check your program.

As the sensor cools, the temperature will eventually drop below a value of 90 hex and the alarm should stop. When it drops below 80 hex, the normal control action should be restored.

Now try the following question which is simply a variation of the previous program.

Question

4.15 Write a program which includes both an 'over-temperature' and an 'under-temperature' alarm. For example, choose a temperature value of 70 hex for the low temperature limit, 80 hex for normal control action, and 90 hex for the high temperature limit.

Summary

Most computer systems contain a large number of subroutines in the monitor ROM. These are often available to the programmer and they provide a simple means of increasing the complexity of programs without a corresponding increase in programming time. Care must be taken that all the necessary input and output conditions are studied so that registers are not modified unexpectedly.

Analogue inputs can be obtained from the applications board by employing the appropriate subroutine. This enables a large range of programs to be designed, the simplest of which is a simulation of a thermostat and its control action.

A simple ON–OFF controller can be further enhanced by including an alarm in the program. This then becomes the type of program which could be used as the basis of a typical industrial controller.

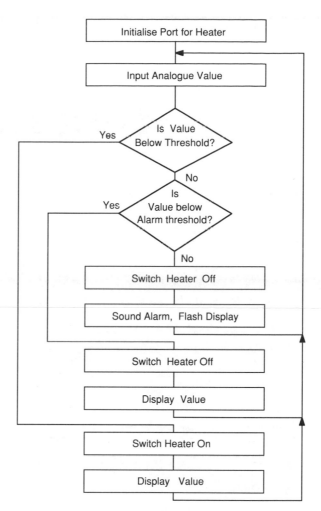

Figure 4.8 Thermostat flow chart with alarm

Sequencing

EQUIPMENT REQUIRED

To complete all of the practical exercises in this chapter, you will need:

(a) Micro-Professor MPF-1B.
(b) Applications board MAB.
(c) The POLSU replacement EPROM set.

You may also wish to use a cross-assembler system, for which you will require a PC Compatible computer and the assembler, editor, linker software and a connecting lead. The programs in the chapter are also available on disk.

PREREQUISITES

Before studying this chapter you should have:

(a) An understanding of the Z80 instruction set including mnemonic representation.
(b) An understanding of the use of subroutines.

(c) An awareness of the principles of digital-to-analogue conversion.
(d) An appreciation of the use of feedback to control simple systems.

These can be achieved by completing the study of the companion volume in the series, *Microelectronics NII*.

5.1 INTRODUCTION

Manufacturing industry is heavily dependent upon sequencers of various forms for its operation. Almost every mass production process involves performing the same operations on a large number of parts or components.

Sequencers have developed over many years from those driven by steam and mechanical pulleys, cams and levers, through basic logic circuits with transistors and then TTL gates to the modern devices with programmable logic controllers or microcomputers.

Different types of sequencer can be constructed with a basic microprocessor, depending upon whether or not feedback control is required.

The most basic sequencer simply outputs data in regular intervals, irrespective of the actions that may or may not be taking place as a result. One example of this type of sequencer is the device that generates analogue voltages of various waveforms via a digital-to-analogue converter. Even if the final waveform is being affected by faults in other circuits or systems, the computer keeps on generating the same outputs.

An improved type of sequencer generates similar outputs, but not necessarily at regular intervals. For example, a pelican crossing controller, once triggered, goes through a fixed series of events and generates the same sequence each time although the time it spends in each state may vary.

The most widely used type of sequencer includes feedback from the process being controlled so that the sequence is modified if it is not correct or the system is forced to wait until all the input conditions have been achieved. This process is vital in industry since it provides safe operation, and the ability to stop a sequence should a fault develop.

In this chapter all three types of sequence are investigated and the type of software required for each one is examined.

PRACTICAL EXERCISE 5.1 – WAVE-FORM GENERATION

Experimental Concepts

Waveform generation is an example of the use of the microprocessor as the most basic sequencer. It requires an accurate digital-to-analogue converter.

In this exercise the task of the computer is to output the digital equivalent of the required analogue voltage at the correct time. Different waveforms can be produced according to the data stored in the computer memory. Consider the system shown in *Figure 5.1*.

Data to be output is stored in the computer memory in a state table, which is output one byte at a time to the output port. Each data value is converted to its analogue equivalent voltage by the digital-to-analogue converter. In the system on the applications board the digital-to-analogue converter is arranged to produce a voltage of approximately:

$$\text{Analogue output} = \text{Binary value} \times 10 \text{ mV}$$

Thus for example, a binary value of 0 0 0 0 0 0 0 1 is equivalent to 0.01 volts and 1 1 1 1 1 1 1 1 is equivalent to 2.55 volts.

When the digital output changes, the analogue output changes very rapidly to the new voltage, typically within 1 microsecond from one value to the next and a smooth output can only be obtained if external smoothing circuits are used. However, on the bar graph display these rapid voltage changes will not be visible, apart from the switch on and off of each individual light.

The computer program is required to output a series of data values from the table with a fixed delay between them. In the flow chart shown in *Figure 5.2* (page 56), a counter is used to hold the number of data values to be output, and a table pointer is used to locate the next number in the table.

This can be translated into the program on page 56. Note that the state table start address has been chosen as 1900 hex. Register B is the counter register which is loaded with the number of values to be output. In the program shown this has been set at 32 (20 hex).

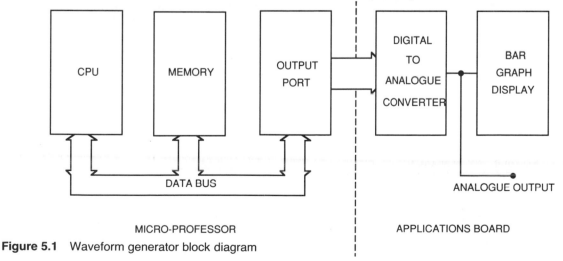

Figure 5.1   Waveform generator block diagram

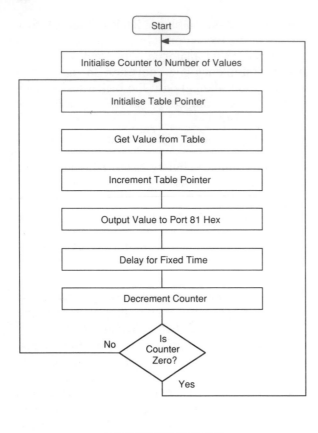

Figure 5.2 Waveform generator flow chart

Program

Address	Hex code		Mnemonic
			ORG 1800H
1800	06 20	START:	LD B,20H
1802	21 00 19		LD HL,1900H
1805	72	LOOP:	LD A,(HL)
1806	23		INC HL
1807	D3 81		OUT (81H),A
1809	11 00 10	DELAY:	LD DE,1000H
180C	1B	DEL:	DEC RE
180D	7A		LD A,D
180E	B3		OR E
180F	C2 0C 18		JP NZ,DEL
1812	05		DEC B
1813	C2 05 18		JP NZ,LOOP
1816	C3 00 18		JP START
			ORG 1900H
1900	N1	TABLE:	DATA
1901	N2		DATA
1902	N3		DATA
–	–		DATA
–	–		
191F	N32		DATA

Procedure

(a) The first task is to calculate the values required for the state table to generate a simple waveform. One waveform which is easy to generate is a stepped sawtooth as shown in *Figure 5.3*.

If a sawtooth with 32 steps is chosen then the values follow the sequence 00, 08, 10, 18, 20, 28 etc.

Write down the 32 values required.

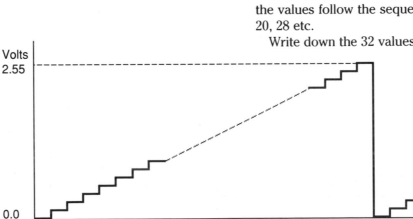

Figure 5.3 Sawtooth waveform

(b) Now enter or download program 1 into the computer starting at address 1800 hex, and the data for the state table starting at address 1900 hex.

(c) On the applications board, make sure that all the switches are OFF, apart from:

BAR GRAPH – ON

(d) Run the program. Write down what happens on the bar graph. Is this what was expected?

Now answer the following questions.

──────── **Questions** ────────

5.1 What is the effect of changing the delay variable loaded into the DE register pair? Try large and small values.

5.2 How can the program be modified so that the waveform only reaches half the maximum output without changing the data values?

5.3 How can the program be modified so that the waveform slopes down instead of up without changing the data values?

Exercises

The following exercises are all based on the same program but use different data tables.

──────── **Questions** ────────

5.4 Work out the state table for the waveform which is shown in *Figure 5.4*.

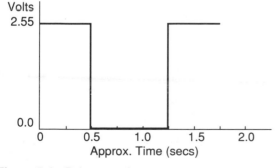

Figure 5.4 Pulse waveform

5.5 It is required to produce a sinusoidal output waveform. Calculate the 36 steps required at 10° intervals and try them in the program.

The sine wave values can be calculated most easily in decimal and then converted into hexadecimal before entry into the computer. Note that the waveform has to be centred around an origin which is half the maximum value. Use the formula given below with values from 0 to 350° for N.

Decimal value $= 128 + 127 \sin N$

───────────────────────────

Many other types of waveform can be generated by changing the state table values, its length and the delay between each state. Experiment with other values and see what interesting waveforms can be produced.

The program given in this section is a general purpose program which can be employed on most waveforms. Whenever waveforms are required in which the values have simple mathematical relationships with one another it is often more convenient to write a program specific to each waveform.

Summary

One of the most basic types of sequencer simply outputs data values with a fixed time delay between them. This can be put to good use as a waveform generator if the computer is equipped with a digital-to-analogue converter. All types of waveforms may be generated with a similar program if a state table in memory contains the sequence of digital values which must be output.

PRACTICAL EXERCISE 5.2 – SEQUENCES WITH VARIABLE DELAYS

Experimental Concepts

The principles of the production of sequences whose time per step is variable has already been explained in the previous exercise. However, this can be expanded slightly here by considering a number of examples. The most widely known sequence of this type is the traffic light sequence. In its simple form this consists of the two sets of lights which govern a cross-roads junction, but far more complex systems exist which cover many other types of junction. For example, the addition of a 'pedestrian' control light changes the sequence table significantly.

Other types of pedestrian crossing are also interesting to analyse. A pelican crossing will be used in this exercise as an example of one that includes a sequence but this does not operate until required to do so.

Sound may be the output of a sequence instead of data on lights. One of the monitor subroutines can be used to generate a simple sequence of sound which could be the basis of a Morse code generator.

At the heart of all of the sequences with variable delays is a delay subroutine which has a minimum delay equal to the smallest time period to be measured. Typically this may be 1 ms, 10 ms, 100 ms or 1 s depending upon the application. The total delay time generated depends upon a parameter passed to the subroutine from the state table.

Program 1

The subroutine below can be used as the basis of the variable delay. Its main loop is a standard delay program with the DE register pair as the counter. If various initial values are put into these registers the minimum time period can be easily adjusted. Parameters are passed to this delay in the C register. The routine has been written to start at address 18A0 hex but this can be changed if necessary.

Address	Hex code		Mnemonic
			ORG 18A0H
18A0	11 n n	VARDEL:	LD DE, nn
18A3	1B	VLOOP:	DEC DE
18A4	7A		LD A,D
18A5	B3		OR E
18A6	C2 A0 18		JP NZ,VLOOP
18A9	0D		DEC C
18AA	C2 A0 18		JP NZ,VARDEL
18AD	C9		RET

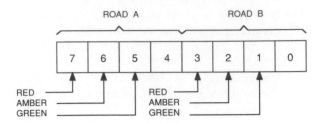

Figure 5.5   Traffic light output connections

Suitable values for nn to obtain different delays are:

1 ms	– 004A hex
10 ms	– 02D8 hex
100 ms	– 1D10 hex
500 ms	– 9150 hex

The 'traffic' lights are connected to the output port 81 as shown in *Figure 5.5*.

Bit 0 is connected to an amber light, but this can be regarded as the 'Pedestrian' control light for road A. This means that it can only allow pedestrians to **cross** road A when its traffic is stopped. Similarly, bit 4 is connected to the 'pedestrian' control light for road B.

Program 2

The main sequencing program is given in mnemonic form below:

	Mnemonic	Comment
	ORG 1800H	
START:	LD B, n	; n = Number of states
	LD HL,1A00H	; State table start
REP:	LD A,(HL)	
	INC HL	
	OUT (81H),A	
	LD C,(HL)	
	INC HL	
	CALL VARDEL	
	DEC B	; Reduce count by 1
	JP NZ,REP	
	JP START	
VARDEL:	– as Program 1	

The state table must also be in memory at address 1A00 hex. Use a basic delay of 500 ms in the following sequences.

Procedure

(A) Traffic light sequence

(a) Construct a state table which will output data for the following sequence to be observed. Remember that in the table the output data is followed in each case by the delay variable so that two bytes comprise each complete state.

RED	1 s
RED	1 s
RED & PEDESTRIAN	4 s
RED	0.5 s
RED & PEDESTRIAN	0.5 s
RED	0.5 s
RED & PEDESTRIAN	0.5 s
RED	0.5 s
RED & PEDESTRIAN	0.5 s
RED	1 s
RED & AMBER	1 s
GREEN	8 s
AMBER	1 s

During the 10 seconds that one set of lights are on red, the other lights go through the red and amber, green, amber sequence.

It will be easier to work out the complete byte of data to output if it is done in two halves, one for each road, and then the halves are put together when one complete sequence has been written down.

(b) Translate program 2 into machine code starting at address 1800 hex and write it down.

(c) Enter the program and state table into the computer in the normal way.

(d) Switch off all the switches on the applications board and then run the program.

If the output sequence is not exactly what was expected, check through the state table again.

(B) Pelican crossing sequence

The pelican crossing sequence is similar in many ways to the traffic lights apart from the fact that it is triggered manually and does not run continuously. The lights on the applications board will have to be interpreted slightly differently, as shown in *Figure 5.6*.

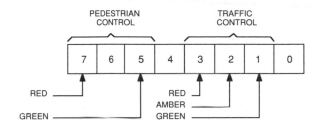

Figure 5.6 Pelican crossing output connections

When the lights have been triggered, the sequence of events is as shown below:

TRAFFIC GREEN and PEDESTRIAN RED	5 s
TRAFFIC AMBER and PEDESTRIAN RED	2 s
TRAFFIC RED and PEDESTRIAN RED	1 s
TRAFFIC RED and PEDESTRIAN GREEN	6 s
TRAFFIC AMBER and PEDESTRIAN GREEN	0.5 s
ALL OFF	0.5 s
TRAFFIC AMBER and PEDESTRIAN GREEN	0.5 s
ALL OFF	0.5 s
TRAFFIC AMBER and PEDESTRIAN GREEN	0.5 s
ALL OFF	0.5 s
TRAFFIC AMBER and PEDESTRIAN GREEN	0.5 s
ALL OFF	0.5 s
TRAFFIC GREEN and PEDESTRIAN RED	0.5 s

The last state remains until the sequence is retriggered.

The simplest means of triggering the sequence is to wait for the [USER] key on the keyboard to be pressed.

The [USER] key is connected directly to bit 6 of port 0. When it is depressed a logic 0 is generated and when it is released a logic 1 is produced.

The program simply has to wait for a logic 0, then a logic 1 on port 0 bit 6. Proceed as follows:

(a) Write out the state table for the pelican crossing with each output state followed by the time code required.

(b) Modify program 1 to include a routine that waits for the key depression of the [USER] key before the main part of the program begins.

(c) Translate the program into machine code and download or enter it into the Micro-Professor memory starting at address 1800 hex. Enter the state table into memory starting at address 1A00 hex.

(d) Run the program.

Now press the [USER] key on the Micro-Professor keyboard and observe the lights on the applications board. Do they behave as expected? If not, check your state table again.

Now answer the following question.

———————— **Question** ————————

5.6 If a subroutine were available that would bleep the speaker for 6 seconds, briefly describe, without writing any programs, how this could be included in the program to make it more realistic.

For example, compare your program with the one at address 0D15 hex, which was built into the POLSU replacement EPROM set.

_____

(C) Sound Sequence

A subroutine exists in the monitor ROM which generates a 1 kHz note. The complete information is given below:

Name	: TONE 1
Address	: 247C hex
Function	: Generate sound at 1 kHz
Input	: Number of cycles in DE pair
Output	: None other than sound
Registers destroyed	: AF, BC, DE

Clearly, 100 cycles (64 hex) represents a time period of 0.1 seconds at 1 kHz.

Morse code consists of bursts of a set frequency with periods of silence between them, so it should be possible to arrange a program to generate the following sequence. It represents the word CODE in Morse code.

C	1 kHz tone	–	0.3 s
	Silence	–	0.1 s
	1 kHz tone	–	0.1 s
	Silence	–	0.1 s
	1 kHz tone	–	0.3 s
	Silence	–	0.1 s
	1 kHz tone	–	0.1 s
	Silence	–	0.3 s
O	1 kHz tone	–	0.3 s
	Silence	–	0.1 s
	1 kHz tone	–	0.3 s
	Silence	–	0.1 s
	1 kHz tone	–	0.3 s
	Silence	–	0.3 s
D	1 kHz tone	–	0.3 s
	Silence	–	0.1 s
	1 kHz tone	–	0.1 s
	Silence	–	0.1 s
	1 kHz tone	–	0.1 s
	Silence	–	0.3 s
E	1 kHz tone	–	0.1 s
	Silence	–	0.5 s

The only real difference between this and the previous sequences is that the data for the tone duration has to be loaded into the DE register pair and therefore occupies 2 bytes in the state table. This is then followed by a single byte for the delay code which will be used in the variable delay subroutine as before to give a period of silence.

For example, the state table for a 0.1 second burst of 1 kHz followed by a 0.1 second silence would be:

1A00	64	Tone
1A01	00	
1A02	01	Silence

0064 hex represents 100 cycles of the 1 kHz tone which last 0.1 seconds.

Program 2 has to be modified as shown below:

Delete the instructions	LD A,(HL)
	INC HL
	OUT (81H),A
Replace them with	LD E,(HL)
	INC HL
	LD D,(HL)
	INC HL

PUSH BC
CALL TONE1
POP BC

Now proceed as follows:

(a) Write down the values required in the state table for the sequence shown above.

(b) Re-write the complete sequencing program with the necessary modifications.

(c) Now assemble the program starting at address 1800 hex.

(d) Enter or download the program into the Micro-Professor, followed by the state table at the correct address.

(e) Execute the program.

How does it sound? The whole sequence should last about 4 seconds and then repeat itself after a longer silence.

Now answer the following questions.

—————— Questions ——————

5.7 What limits the length of Morse code message which could be sent using the program you have written? How could this be extended?

5.8 What is the lowest number of cycles of 1 kHz which gives a perceptible note from the TONE 1 subroutine?

Summary

Many applications exist that require a sequence of events to take place but the time per event is not fixed. These include systems such as traffic lights, pelican crossings and Morse code generators. A microcomputer can be used for such sequences when it is equipped with software that reads all its variables from a state table. Each entry in the state table consists of two values, the first that represents the output data and the second that represents a duration code. Sequences can be made very long, if necessary, by equipping the computer with sufficient memory.

PRACTICAL EXERCISE 5.3 – CONDITIONAL SEQUENCES

Experimental Concepts

Conditional sequencing is a widely used technique in an industrial environment. It provides a means of 'feedback' from the devices controlled by the sequencer so that actions do not take place until the previous ones are complete. This is very important from both a safety and a reliability point of view. The computer software and state tables are very similar to those already encountered in the previous exercises in this chapter.

Probably the most widely known sequence of this type is the domestic automatic washing machine. In this exercise it will be possible to produce a simulation of the sequence using the switches and lights as the inputs and outputs.

Washing Machine Simulation Connections

Port 81 hex drives the outputs as follows, all of which are active at logic 1:

Bit 0	LOCK DOOR
1	START TIMER
2	PUMP ON
3	COLD VALVE OPEN
4	HOT VALVE OPEN
5	HEATER ON
6	DRUM FAST
7	DRUM SLOW

Port 80 hex receives the inputs, which are also active at logic 1:

Bit 0	TEMPERATURE CORRECT
1	LEVEL FULL
2	LEVEL EMPTY
3	2 MINUTE TIME COMPLETE
4	5 MINUTE TIME COMPLETE
5	DOOR CLOSED
6	UNUSED – LOGIC 0
7	UNUSED – LOGIC 0

Sequence required A simplified wash rinse and spin sequence and associated flow chart (*Figure 5.7*) are shown overleaf. Note that the door must remain closed and locked throughout the cycle:

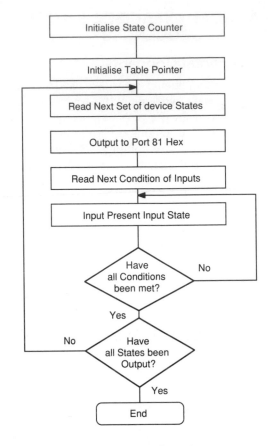

Figure 5.7   Washing machine flow chart

Step	Action	Terminating Condition
1.	Wait	Door closed (and level empty)
2.	Lock door	
	Hot valve open	Level full
3.	Heater on	Temperature correct
4.	Drum slow	
	Start timer	5 minute time complete
5.	Pump on	Level empty
6.	Cold valve open	Level full
7.	Drum slow	
	Start timer	2 minute time complete
8.	Pump on	Level empty
9.	Pump on	2 minute time complete
	Drum fast	(and level empty)
	Start timer	
10.	Unlock door	All inputs at 0 except level empty.

Procedure

(a) The first task is to translate the **sequence required** into a series of logic states and set up the state table in memory.

Construct a similar table for the washing machine sequence to those constructed previously but note that when this is entered into memory, the controlled device state must come in the first byte of each state followed by a byte containing the conditional input states (terminating condition). Work each state out in binary then change the binary values into hexadecimal.

Check your state table carefully.

(b) Now write a suitable program for the sequencer based on the flow chart. Start the program at address 1800 hex as usual.

(c) Enter the program and state table into the Micro-Professor starting at addresses 1800 hex and 1A00 hex respectively.

(d) Put all the switches of port 80H at logic 0 and then run the program.

Initially nothing should happen!

(e) Set up the required input conditions for the first state on the port 80H switches and check that the next output state is generated.

Proceed to change the port 80H switches to each of the required input states and make sure that on each occasion the next state is generated.

When all the inputs have been entered successfully, the program should HALT.

Now answer the following questions.

——— Questions ———

5.9 The unused inputs in the previous program were assumed to be at a logic 0. However, if they were not at 0 the program would not continue. How could these bits be ignored in the program?

5.10 Give two other applications for sequencing programs of this type.

5.11 A program is required which will act as a computer controlled combination lock. It requires the following numbers to be entered on the port 80H switches in the correct order.

81, 36, F2, 9A, C1, D5, 0E and 74

After each state has been correctly input one of the port 81 lights comes on starting at bit 0 and working towards bit 7. (This would be omitted in a real lock, but it is included here to make the program testing easier.)

Work out a suitable state table and try the same program used for the washing machine with the new table.

Summary

A washing machine and electronic combination lock are examples of systems that require a sequence of events to take place in a given order. Each step in the sequence must take place before the next can be started. There are many such devices in industry particularly in the manufacturing and automatic testing fields. Most systems can be computer controlled and the software used in this experiment could be used in many of them. The only real difference between controllers is the state table which governs the actions that take place and the required responses.

Peripheral control

EQUIPMENT REQUIRED

To complete all of the practical exercises in this chapter, you will need:

(a) Micro-Professor MPF-1B.
(b) Applications board MAB.
(c) The POLSU replacement EPROM set.

You may also like to use a cross-assembler on an IBM PC and download the HEX files to the Micro-Professor RAM. This will require a suitable assembler, editor and linker with a connecting lead.

The programs are also available on disk to save a lot of typing.

PREREQUISITES

Before studying this chapter, you should have:

(a) An appreciation of the use of different formats for number representation in microcomputers.

(b) An awareness of the use of programmable input/ output devices for peripheral interfacing.
(c) An appreciation of the use of buffers and driver circuits in digital electronics.

These may be achieved when the accompanying volume in the series, *Microelectronics NII* has been completed, and Chapters 1, 2 and 6 of *Microelectronics NIII* have been studied.

6.1 INTRODUCTION

Almost every microcomputer has a keyboard which may have as few as 16 keys or as many as 102. For large keyboards it is generally necessary to employ a dedicated integrated circuit to monitor the keys continuously and detect every key depression. The keyboard encoder chip also assigns the correct code to each key so that the processor has very little work to do other than accept the keyboard data.

When systems employ small keyboards it is often more convenient to allow the microprocessor to scan the keys for every depression and then assign the appropriate code to the key. This is a technique known as '**software encoding**' and is widely used. In the Micro-Professor the keyboard uses this technique for all its data entry.

Similarly, displays in large systems may be controlled by special circuits known as '**display controllers**'. These are relatively expensive, so small systems frequently dispense with the need for them by employing another software technique known as '**display multiplexing**'. Using this technique, the 7-segment displays in a system are sent data very rapidly, but only to one display digit at a time. This happens so fast that with the persistence of vision of the human eye, the displays appear to be on continuously.

All data sent to 7-segment displays must be in the correct format to illuminate the appropriate segments of the display. Unfortunately, this format is not usable for any other purpose within the system, so all data must be converted to it prior to display. Therefore routines are frequently needed to convert from binary, ASCII or BCD into 7-segment format.

This chapter examines some of the more practical aspects of real microcomputer interfacing.

PRACTICAL EXERCISE 6.1 – KEYBOARD SCANNING

Experimental Concepts

The Micro-Professor keyboard has 36 keys. Although the physical layout is on a 9×4 grid the electrical connections are slightly different. The four keys in the left-hand column of the keyboard each have separate connections to the system. The remaining 32 keys are arranged on a 6×6 matrix with four blank positions as shown on the circuit diagram. Each key is connected to the cross-point of a 'vertical column' and a 'horizontal row' conductor. When a key is depressed it connects one of the columns to one of the rows. When there are no keys depressed, there are no connections between the rows and columns, and the row lines are connected to the 5 volt supply via 'pull-up' resistors. The 'pull-up' resistors ensure that a logic 1 will be present on the row lines when no keys are depressed.

On the circuit diagram (*Figure 6.1*) (overleaf), connections PC0–PC5 are arranged to be outputs. They are the lower 6 bits of port 02. Connections PA0–PA5 are inputs. They are the lower 6 bits of port 00.

The keyboard scanning software operates by checking each key in turn. A single logic 0 is sent to one of the outputs PC0–PC5 with all the other outputs at logic 1. Initially the logic 0 is sent to PC0.

Each input line PA0–PA5 is then tested to see if it is connected to a logic 0 which would indicate that one of the keys in the right-hand column was depressed. After each test is made, a counter is incremented so that it keeps track of which key is being tested at any moment. The counter assigns a unique number to each key. Because this number is determined by the program the keyboard is said to be 'software encoded'.

When the right-hand column has been checked, the data at the output port is arranged so that the logic 0 moves along from PC0 to PC1. This then allows the second column of keys from the right to be tested. The process then continues with the logic 0 moving along the columns and each row of keys being checked until the whole keyboard has been tested. If at any time a key is found to be depressed the value in the counter is noted and appropriate action can be taken.

The flow chart for the keyboard scanning program is shown in *Figure 6.2* (page 67). Three counters are required, one to keep a record of the key being checked and another to count down from 6 so that only 6 rows are checked in each column. The third also counts down from 6 so that only 6 columns are checked. Some means of checking whether a key has been depressed must also be included and this is the 'key detector', which in practice could be a CPU flag or one bit of a register.

The whole program has been written as a subroutine so that it can be used as part of other programs.

Program

The flow chart has been translated into the following program using the CPU registers below:

Register B – Row counter
Register C – Key counter
Register D – Row data
Register E – Column output data
Register H – Column counter
Register L – Key detector

Memory Address 1A00 hex – key code of any key pressed.

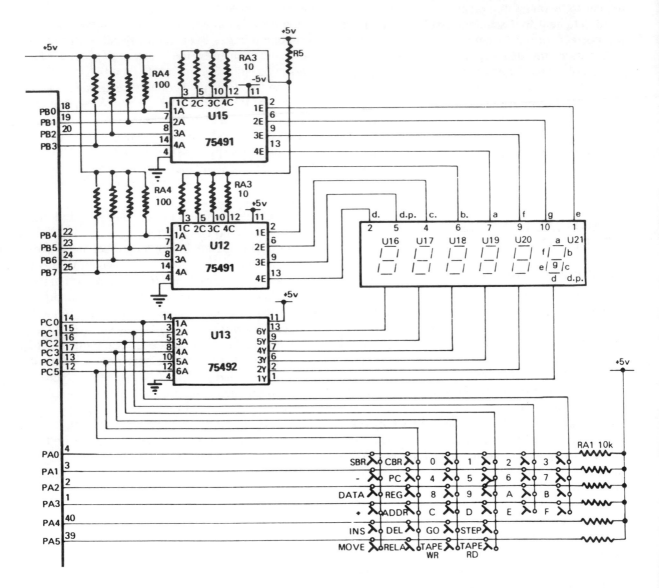

Figure 6.1   Keyboard and display circuit

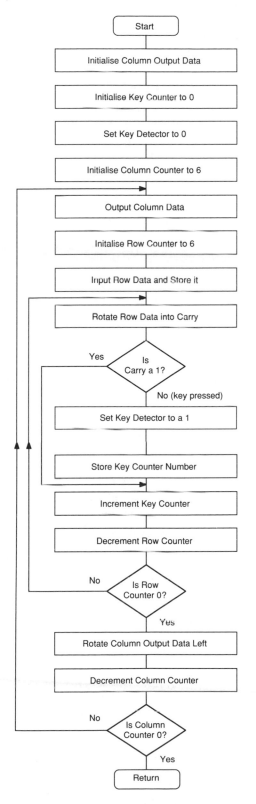

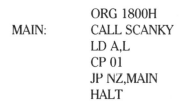

	Mnemonic		Comment
	ORG 1900H	;	Subroutine origin
SCANKY:	LD E,0FEH	;	1 1 1 1 1 1 1 0 in binary
	LD C,0	;	Key counter is 0
	LD L,0	;	No keys detected
	LD H,6	;	6 columns
KCOL:	LD A,E		
	OUT (02),A	;	Output column data
	NOP	;	No operation – for later use
	LD B,6	;	6 rows
	IN A,(00)	;	Input row data
	LD D,A	;	Store it in D
KROW:	RR D	;	Rotate bit 0 of D into carry
	JP C,NOKEY	;	Test carry
	LD L,01	;	Key detected
	LD A,C		
	LD (1A00H),A	;	Store key number in memory
NOKEY:	INC C	;	Increment key counter
	DEC B	;	Decrement row counter
	JP NZ,KROW	;	Test next row
	RLC E	;	Rotate column data left
	DEC H	;	Decrement column count
	JP NZ,KCOL	;	Test next column
	RET		

Procedure

(a) The first task is to assemble the program into hexadecimal code and enter it into the Micro-Professor. Since this is a subroutine, start it at address 1900 hex. This will allow any 'main' programs to start at 1800 hex. Check the program carefully since it is relatively long.

(b) A very simple main program is now required which will CALL the scanning routine and halt whenever a key is pressed. The program below should suffice:

```
        ORG 1800H
MAIN:   CALL SCANKY
        LD A,L
        CP 01
        JP NZ,MAIN
        HALT
```

Assemble this and enter or download it into the Micro-Professor starting at address 1800 hex.

(c) Before running the program at full speed, it will be possible to investigate its operation using the logic probe. If this is not already connected, put the probe lead into the socket on the applications board.

Figure 6.2 Keyboard scanning flow chart

To investigate the program operation with the logic probe it will be necessary to STOP the program at a certain point and then check the logic states on the keyboard lines. However, if the program is stopped in the normal way using a breakpoint, the monitor program takes over the control of the keyboard and scans them waiting for a key depression. This would totally mask any signals which are generated by the program.

There is another way of stopping a program, however, and that is to use a HALT instruction.

Therefore, in the program, REPLACE the NOP instruction with a HALT instruction (Code 76). This will stop the program after the output instruction and allow the keyboard lines to be examined.

(d) Now find pin 14 of U14 and hold the logic probe on it. Both the HIGH and LOW lights should be ON and the PULSE light flashing which indicates that the keyboard is being checked by the monitor program.

(e) Execute the program at address 1800 hex. The HALT light should come on almost immediately indicating that the first part of the program has been executed and the HALT instruction has been encountered.

Now check the outputs on outputs PC0 to PC5 of U14 with the logic probe. Be careful to note that the pin numbers do not follow in order!

If the program has been entered correctly there should be a logic 0 on pin 14 and a logic 1 on all the other outputs. Enter the results on the chart below.

(f) While the scanning program is stopped, put the logic probe onto pin 4 of U14. This is one of the input lines from the keyboard. If the key in the top right-hand corner of the matrix is pressed, pin 4 should go to a logic 0. Confirm that this is the [3] key. Also check pins 3, 2 and 1 and confirm which keys are in the right-hand column of the matrix.

(g) Now proceed with the rest of the program as follows:
Press [MONI] then [GO] to continue with the

next program loop. The HALT light should come on again since the rest of the program will have been executed. Check the outputs PC0 to PC5 with the logic probe and enter them on the chart below.

(h) Continue to execute the program one loop at a time by pressing [MONI] then [GO] after each time it HALTS and complete the following chart. Do not press any keys other than those indicated while doing this.

Number of passes through the program loop	PC5 (12)	PC4 (13)	PC3 (17)	PC2 (16)	PC1 (15)	PC0 (14)
1						
2						
3						
4						
5						
6						
7						
8						

If the outputs obtained are not what were expected at any point, recheck the program carefully.

(i) If the program is now scanning the keyboard as expected it can be checked for key detection by running it at full speed.

Replace the HALT instruction which was changed earlier with a NOP instruction (00 hex). Now execute the program. Press the [GO] key quickly or the system will HALT since it detects the [GO] key immediately.

Press any one of the keys (apart from those in the left-hand column). The program should now HALT.

(j) Press either [RESET] or [MONI] to restore monitor control and examine the contents of address 1A00 hex. This is where the program stores the key number of the key which was pressed. It should be a number between 00 and 23 hex.

(k) Run the program again and press a different key. Check that the key code produced is different.

──────── **Questions** ────────

6.1 What are the key codes for the following keys on the keyboard?

9, A, STEP, TAPE WR, MOVE

6.2 On the circuit diagram, four keys are missing from the matrix but the program still checks their key positions. What will the key codes be for these missing keys?

───────────────────────────────

In the following questions make use of the keyboard scanning subroutine and only require changes to the main program which calls it.

──────── **Questions** ────────

6.3 The Micro-Professor bleeps the loudspeaker every time a key is pressed. Write a program which makes use of the TONE1 subroutine explained in Exercise 5.2 to bleep the loudspeaker for 0.1 seconds when a key is pressed.

6.4 Modify the previous program so that the loudspeaker bleeps only when the [STEP] key is pressed.

───────────────────────────────

Summary

Keyboards are very important computer peripherals. Generally in small systems they are connected as a matrix of switches directly to an output and input port. Each switch in the keyboard is checked in turn by the software which is known as a keyboard scanning program. Whenever a key depression is detected the keyboard scanning program can identify the key by a unique number known as the key code. Appropriate action may then be taken depending upon the key pressed.

The monitor program normally performs all the keyboard scanning required so that the user need not be aware of the complex processes involved.

PRACTICAL EXERCISE 6.2 – DISPLAY MULTIPLEXING

Experimental Concepts

In the Micro-Professor computer, the display and keyboard are both connected to U14 which is an 8255 chip containing three independent 8-bit ports. Each port C pin is used to scan both the keyboard and the display. Although the keyboard is connected directly to the ports the display requires some extra buffer chips to provide the high currents needed to generate sufficient light output. On the circuit diagram these buffers are the two 75491 chips and 75492 which are designated U12, U15 and U13 respectively on the circuit board (*Figure 6.4*) (page 70). The two 75491 chips are known as the **segment drivers** and the 75492 chip is the **digit driver**.

The display contains six 7-segment displays. Each one is connected in 'common cathode' configuration which is shown in *Figure 6.3*. The segments are indicated with the letters a to g and the decimal point is p.

Any segment can be illuminated by sending a logic 1 to the appropriate **segment** drive pin and a

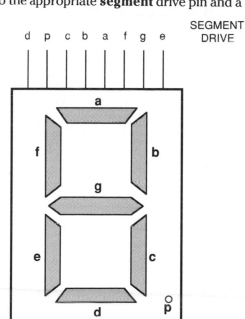

Figure 6.3 Seven segment display

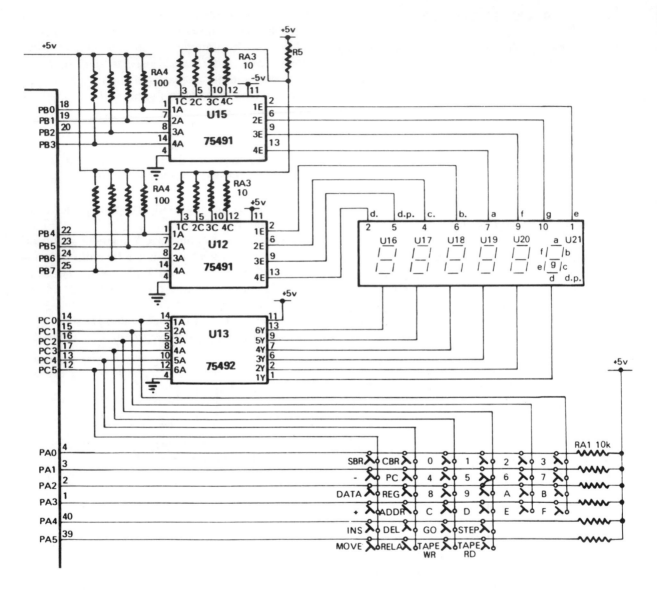

Figure 6.4   Keyboard and display circuit

logic 0 to the **digit** drive. In the complete display all the segment drive lines are connected in **parallel**, so that whatever code is sent out from port B of U14, PB0–PB7, will go to **all** the six displays.

Each digit drive line D_1 to D_6, however, has an independent connection via the 75492 to port C of U14, PC0–PC5. Note that although a logic 0 is required to turn any digit on, the 75492 chip contains an **inverter**. This means that to illuminate any digit a logic 1 must be sent from the port.

Only one digit is illuminated at a time. For each digit, the appropriate code is sent to the segment drive lines then the digit line turns one of the displays ON. There is a short delay and then the digit line turns the display OFF. Each digit is illuminated in turn but this happens so quickly that they all appear to be on simultaneously. This is described in *Figure 6.5*.

Because there are six displays, the six characters which are to be displayed must be stored somewhere in the computer system. This could be in the registers but a more usual technique is to reserve **six** bytes of memory for them. This reserved area of memory is known as the **display buffer**. One of the 16 bit registers can be used as a **pointer** to the data in the buffer.

Figure 6.5 and the following program assume that a display buffer exists in memory with the right-hand digit of the display stored in the first memory address and the left-hand digit stored in the last memory address.

Program

The display multiplexing program uses the CPU registers as follows:

Register pair	BC	– Delay variable
Register	E	– Digit code
Register	H	– Digit counter
Index register	IX	– Display buffer pointer
Port	01	– Segment display
Port	02	– Digit data
Memory addresses		
1A00–1A05 hex		– Display buffer

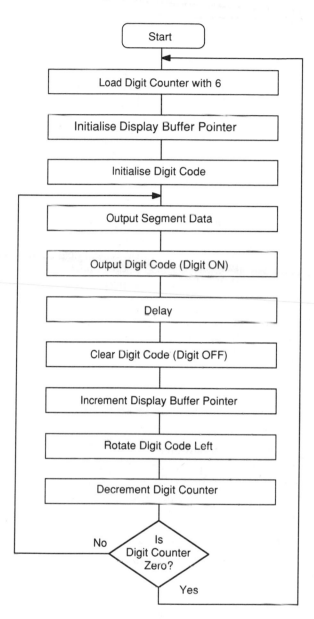

Figure 6.5 Multiplexed display flow chart

Note that bits 6 and 7 of the digit output port (PC6 and PC7) must be kept at logic 1 throughout the program. This is because they are connected to other parts of the computer circuit. An extra instruction has been included to achieve this, i.e. OR 0C0 hex.

	Mnemonic	Comment
	ORG 1800H	
MPLEX:	LD H,6	; Digit counter
	LD IX,1A00H	; Display buffer pointer
	LD E,01	; Right-hand digit code
DIG:	LD A,(IX+0)	
	OUT (01),A	; Output segment data
	LD A,E	; Get digit data
	OR 0C0H	; Bits 6 and 7 to logic 1
	OUT (02),A	; Turn digit ON
	CALL DELAY	
	LD A,0C0H	; Clear bits 0–5 of A
	OUT (02),A	; Turn digit OFF
	INC IX	; Move buffer pointer
	RLC E	; Rotate digit code
	DEC H	; Decrement digit count
	JP NZ,DIG	; Loop for next digit
	JP MPLEX	; Repeat
	ORG 1830H	
DELAY:	LD BC,0000	; Initial delay value
DEL:	DEC BC	
	LD A,B	
	OR C	
	JP NZ,DEL	
	RET	

Note that the initial delay value given will provide about half a second delay per digit. This will allow the multiplexing program to be seen at a very slow rate. When the program is operating correctly it will be possible to speed it up by reducing the delay time.

Procedure

(A) Multiplexing program

(a) The first task is to decide upon a message to display and to work out which segments need to be illuminated for each character.

Suppose the message is 654321.

Complete the table below for each character.

Display	Segments	d	p	c	b	a	f	g	e	Hex code
1	b,c	0	0	1	1	0	0	0	0	30
2	a,b,d,e,h	1	0	0	1	1	0	1	1	9B
3										
4										
5										
6										

(b) Convert the multiplexing program to machine code starting at address 1800 hex. Do not forget the delay subroutine which can either follow directly after the main program or be placed at any convenient memory address.

(c) Enter the program into memory at address 1800 hex. Enter the hex code for the display message at address 1A00 hex. Remember that the right-hand digit (1) goes into the first memory address (1A00 hex).

(d) Now execute the program.

If the program is working correctly the right-hand digit should display a 1 for about half a second and then the next display should show a 2 for the same length of time. Each display should switch on in turn until the complete message is displayed. The process then repeats itself.

(e) If the program works correctly, stop it and reduce the delay time between digits.

Try a range of values between 8000 hex and 0001 hex in the delay subroutine.

Questions

6.5 What range of delay values give an acceptable display without any noticeable flicker?

6.6 What happens when the delay time is very short?

(f) Now try some other messages, for example:

POLSU
or HELP US

Question

6.7 How can the program be modified so that the display is scanned from left to right instead of right to left?

(B) Using the monitor display subroutine
Sending information to the display is a very important aspect of the computer's function and naturally the monitor program contains routines that allow it to perform this. Fortunately these are

written as subroutines so that the user can also have access to them.

In the monitor program, the display scanning routine is **combined** with the keyboard scanning routine. This means that every time the display is scanned the keyboard is also checked for inputs. However, it is possible to use either the display or keyboard parts of the program independently by ignoring the requirements of the unwanted part.

The full subroutine specification is given below.

Name : SCAN1
Address : 0624 hex
Function : Scan keyboard and display 1 cycle from right to left. Execution time is 9.97 ms.
Input : IX points to the start of the display buffer
Output : (a) If no key is pressed then carry flag = 1
(b) If a key is pressed then carry flag = 0 and A contains the key code.
Registers destroyed : AF, AF', BC', DE'
Note : (a) 6 bytes are required to store the seven segment patterns to be displayed.
(b) IX points to the right-most digit.
IX + 5 points to the left-most digit.

To use only the display part of the program simply load IX with the start address of the display buffer and then CALL the subroutine. To use only the keyboard scanning part, load IX with address 07A5 hex which will **blank** the display. Both parts of the program can be used simultaneously by choosing a suitable display buffer.

Proceed as follows:

(a) First, confirm the operation of this monitor subroutine by writing a short main program which calls it repeatedly.

```
        ORG 1800H
MAIN:   LD IX,1A00H
        CALL SCAN1
        JP MAIN
```

Before running the program make sure that the display buffer contains some sensible data.

(b) Displays can be made more interesting if they are made to flash. This can be achieved very simply by calling the SCAN1 routine a fixed number of times to display a message and then calling it a fixed number of times with a blank display.

The SCAN1 routine may be called 50 times for example, to give a display lasting about half a second, by using the program below:

```
         LD B,32H
HLFSEC:  CALL SCAN1
         DEC B
         JP NZ,HLFSEC
```

Now answer the following question.

Question

6.8 Write a program which will flash the message HELLO on the five right-hand displays. The display should stay on for about half a second each time.

(c) Moving displays can be produced almost as easily as flashing ones. To make the display appear to step across from one position to the next, the method is simply to change the address of the display buffer (the value in IX) when each display has been on for say, 0.4 seconds. A counter is also required to make the program stop at the end of the message and go back to be beginning.

Consider the program below:

	Mnemonic	Comment
	ORG 1800H	
MOVER:	LD C,2AH	; C holds number of digits
	LD IX,1A2AH	; Display buffer start
MOVE:	LD B,28H	
MOVE:	CALL SCAN1	; Scan for 0.4 seconds.
	DEC B	
	JP NZ,MOV1	
	DEC IX	; Move pointer
	DEC C	; Decrement counter
	JP NZ,MOVE	; Test for end of message
	JP MOVER	; Repeat message

The values in IX and register C have been arranged for the message below.

Address	Data							
1A00	00	00	00	00	00	00	87	AE
1A08	8F	A7	00	AE	30	00	03	A3
1A10	87	AE	3F	8D	00	87	A1	A7
1A18	00	B3	A3	A3	AD	00	AE	30
1A20	00	85	30	BD	00	85	85	8F
1A28	37	AE	00	00	00	00	00	00

Convert the previous program into machine code and enter it into the Micro-Professor starting at address 1800 hex. Enter the message above into the addresses shown.

Now run the program and write down the message.

Try the following questions.

———— Questions ————

6.9 Write a program which will display the message:

COFFEE or tEA?

stepping across the display from right to left.

6.10 Write a program which will flash the word HELP on the display until the [STEP] key is pressed.

Summary

A group of seven segment displays provides a convenient method of data output from small computers. The most common method of connection is to employ a multiplexing circuit and its associated software. Multiplexing reduces the number of connections required between the computer and display but the driving software is relatively complex. It consists of a program which switches on each display for a short period of time but does this so quickly that they all appear to be on simultaneously.

A monitor program generally contains routines which allow the computer user to display messages. By employing the monitor routines it is very easy to produce a variety of display types such as those that flash or move across from one side to the other.

PRACTICAL EXERCISE 6.3 – NUMBER FORMAT CONVERSION

(A) Binary to BCD Conversion

Experimental concepts In computer applications that require a display of numbers, it is generally necessary to provide this in decimal form. Since most computer calculations are carried out in binary, then some means of converting from binary to decimal format must be provided. Decimal numbers are usually stored in a format known as **binary coded decimal (BCD)**.

BCD format uses four bits to represent a single decimal digit. Thus each byte can store two BCD digits. For example the decimal number 75 could be represented as:

$$0\ 1\ 1\ 1 \qquad 0\ 1\ 0\ 1$$
$$7 \qquad\qquad 5$$

The highest number than can be represented in BCD format in a single byte is 99.

A single binary byte can hold numbers up to 255. This means that if an 8-bit binary number is to be converted into BCD format, then two bytes will be required to store the result, 12 bits of which will be used.

For example:

$$1\ 1\ 1\ 1\ 1\ 1\ 1\ 0 = 0\ 0\ 0\ 0 \quad 0\ 0\ 1\ 0 \quad 0\ 1\ 0\ 1 \quad 0\ 1\ 0\ 0$$
$$254 \qquad\qquad 2 \qquad\quad 5 \qquad\quad 4$$

The conversion process may be carried out in a number of ways but the one that follows is probably one of the simplest. It makes use of the DAA (decimal adjust accumulator) instruction which correct the results of binary calculations so that they become true BCD numbers.

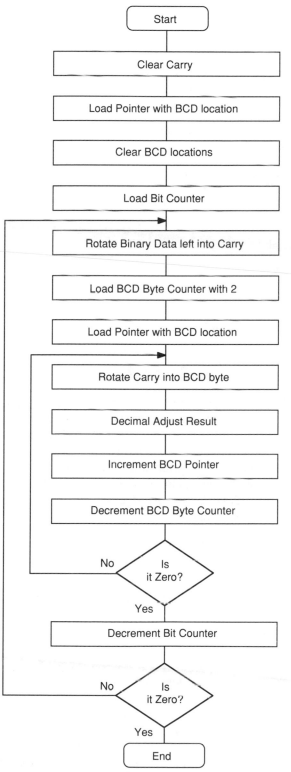

Figure 6.6 Binary to BCD flow chart

The binary number is shifted left one bit at a time into the locations which will hold the BCD number. After each step, the BCD number is decimal adjusted. Eight shifts are required to complete the conversion.

The technique can be illustrated by considering the following example which converts the binary number 1 0 0 0 0 0 0 1 into BCD 129.

Step		BCD	Binary	
		0 0 0 0	0 0 0 0 0 0 0 0	1 0 0 0 0 0 0 1
1.	Shift	0 0 0 0	0 0 0 0 0 0 0 1	0 0 0 0 0 0 1 0
	DAA	0 0 0 0	0 0 0 0 0 0 0 1	0 0 0 0 0 0 1 0
2.	Shift	0 0 0 0	0 0 0 0 0 0 1 0	0 0 0 0 0 1 0 0
	DAA	0 0 0 0	0 0 0 0 0 0 1 0	0 0 0 0 0 1 0 0
3.	Shift	0 0 0 0	0 0 0 0 0 1 0 0	0 0 0 0 1 0 0 0
	DAA	0 0 0 0	0 0 0 0 0 1 0 0	0 0 0 0 1 0 0 0
4.	Shift	0 0 0 0	0 0 0 0 1 0 0 0	0 0 0 1 0 0 0 0
	DAA	0 0 0 0	0 0 0 0 1 0 0 0	0 0 0 1 0 0 0 0
5.	Shift	0 0 0 0	0 0 0 1 0 0 0 0	0 0 1 0 0 0 0 0
	DAA	0 0 0 0	0 0 0 1 0 1 1 0	0 0 1 0 0 0 0 0
6.	Shift	0 0 0 0	0 0 1 0 1 1 0 0	0 1 0 0 0 0 0 0
	DAA	0 0 0 0	0 0 1 1 0 0 1 0	0 1 0 0 0 0 0 0
7.	Shift	0 0 0 0	0 1 1 0 0 1 0 0	1 0 0 0 0 0 0 0
	DAA	0 0 0 0	0 1 1 0 0 1 0 0	1 0 0 0 0 0 0 0
8.	Shift	0 0 0 0	1 1 0 0 1 0 0 1	0 0 0 0 0 0 0 0
	DAA	0 0 0 1	0 0 1 0 1 0 0 1	0 0 0 0 0 0 0 0
		1	2 9	

Notice how the DAA instruction can be used to correct the calculation in steps 5, 6 and 8. It does this by checking the carry flag, the half-carry flag (carry between bits 3 and 4) and the magnitude of the numbers in each half of the accumulator. If a number has to be corrected by the DAA instruction, a 6 is added to the relevant half of the accumulator.

Figure 6.6 assumes that the binary number to be converted to BCD is stored initially in a register and that its BCD equivalent will be stored in two adjacent memory locations pointed to by a register pair.

Program 1

On entry to the subroutine below, the binary number to be converted must be in the C register. The BCD result is placed in addresses 1A00 and 1A01H.

Address	Hex code		Mnemonics		
			ORG 1940H		
1940	AF	BINBCD:	XOR A		
1941	21 00 1A		LD HL,1A00H	;	BCD
1944	77		LD(HL),A	;	CLEAR 1A00H
1945	23		INC HL		
1946	77		LD (HL),A	;	CLEAR 1A01H
1947	06 08		LD B,8	;	BIT COUNTER
1949	CB 11	LOOP:	RL C		
194B	16 02		LD D,02	;	BYTE COUNTER
194D	21 00 1A		LD HL,1A00H		
1950	7E	BCDADJ:	LD A,(HL)		
1951	8F		ADC A,A	;	SHIFT LEFT
1952	27		DAA		
1953	77		LD (HL),A		
1954	23		INC HL		
1955	15		DEC D		
1956	C2 50 19		JP NZ,BCDADJ		
1959	05		DEC B		
195A	C2 49 19		JP NZ,LOOP		
195D	C9		RET		

Procedure

(a) Since the program above has been written as a subroutine, the only way to test it is to **call** it from a **main** program having set up the C register as required.

Start by writing a short main program which loads C with the binary number to be converted and then calls the subroutine. End the program with a **halt** instruction.

(b) Enter the main program and the subroutine into the Micro-Professor starting at addresses 1800H and 1940H respectively.

(c) Run the main program. When it halts press [MONI] to regain control and then examine the memory locations 1A00H and 1A01H to find out if it has worked or not.

When it has performed one conversion successfully, try the following:

(i) 1 0 1 0 0 1 0 1.
(ii) 1 0 0 1 0 0 1 0.
(iii) 1 1 1 1 1 1 1 1.
(iv) 1 1 0 0 1 1 0 0.

Check the results by performing the same binary to decimal calculation on paper.

——— Question ———

6.11 The previous program may be used as the basis for a larger conversion program for multi-byte binary numbers. Modify it so that the 16 bit number held in the BC register pair is converted into BCD format and is stored in addresses 1A00H, 1A01H and 1A02H. Use register E as the bit counter instead of B.

(B) BCD to 7-segment Conversion

Experimental concepts A 7-segment display contains seven independent light emitting diodes which are arranged in the form of a figure 8 and an additional separate diode for the decimal point. The segments are lettered a–g as shown in *Figure 6.7*.

Each diode is normally connected to a separate bit of an output port. In the Micro-Professor, the arrangement is as shown in *Figure 6.8*. When a logic '1' is sent to the port, the corresponding segment is illuminated.

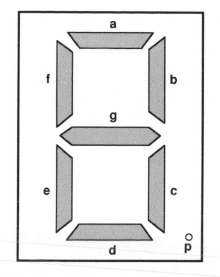

Figure 6.7 Micro-Professor 7-segment display

Figure 6.9 BCD '5' in 7-segment format

simple matter to work out the codes for other letters or symbols.

When numbers are stored in BCD format, each byte can hold two digits. These must be converted separately into two 7-segment codes and stored in their own memory location (*Figure 6.10*).

Since the Micro-Professor has six displays, there could be three memory bytes required to store the BCD data and six bytes to store the 7-segment decoded data. Special reserved areas of memory are often referred to as **buffers** (not to be confused with the hardware type of buffer).

Before a number can be sent to the port and thus appear as a digit on the display, it will generally have to be converted from either binary or BCD format into the correct 7-segment display code.

For example, if the number 5 is held in the lower half of a register in BCD format, this must be converted to 7-segment format code AE (*Figure 6.9*).

The correct 7-segment codes for all the numbers are shown in *Table 6.1*. It is a relatively

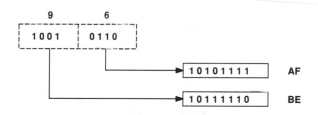

Figure 6.10 '96' in 7-segment code

The conversion process is a simple **look-up table**. Each half of the 8-bit BCD number has to be dealt with separately. A subroutine which converts the number in the lower half of the accumulator into 7-segment format is illustrated in *Figure 6.11*.

BIT	7	6	5	4	3	2	1	0
SEGMENT	d	p	c	b	a	f	g	e

Figure 6.8 7-segment connections

Table 6.1 7-segment code table

Number	7-segment code
0	BD
1	30
2	9B
3	BA
4	36
5	AE
6	AF
7	38
8	BF
9	BE

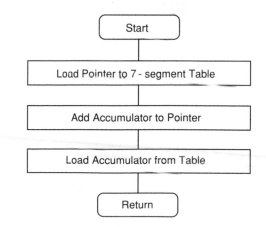

Figure 6.11 BCD to 7-segment flow chart

The number to be converted occupies bits 0–3 of the accumulator. It is assumed that bits 4–7 of the accumulator all contain 0.

A program that converts a two-digit BCD number into two bytes of 7-segment decoded data is illustrated below. The conversion subroutine HEX 7 is based on *Figure 6.11*.

Registers are used as follows:

DE points to BCD data in memory
HL points to 7-segment destination buffer

It is assumed that the BCD data is initially stored in address 1A00H and that the decoded data will be placed in addresses 1A10H and 1A00H respectively. Both the DE pair and the HL pair are incremented near the end of the program so that they are pointing at the next memory locations. This will allow for easy development into a more general routine.

Program 2

Address	Hex code		Mnemonics		
			ORG 1960H		
1960	11 00 1A	BCD7SG:	LD DE,1A00H	;	SOURCE
1963	21 10 1A		LD HL,1A10H	;	DESTINATION
1966	A1	BCD1:	LD A,(DE)		
1967	E6 0F		AND 0FH	;	MASK MSB
1969	CD 80 19		CALL HEX7	;	CONVERT
196C	77		LD (HL),A	;	STORE
196D	23		INC HL		
196E	1A		LD A,(DE)		
196F	0F		RRCA	;	SHIFT TO
1970	0F		RRCA	;	GET MSB
1971	0F		RRCA		
1972	0F		RRCA		
1973	E6 0F		AND 0FH		
1975	CD 80 19		CALL HEX7	;	CONVERT
1978	77		LD (HL),A		
1979	13		INC DE		
197A	23		INC HL		
197B	76		HALT		
			ORG 1980H		
1980	E5	HEX7:	PUSH HL		
1981	21 90 19		LD HL,TABLE		
1984	85		ADD A,L		
1985	6F		LD L,A	;	MOVE POINTER
1986	7E		LD A,(HL)		
1987	E1		POP HL		
1988	C9		RET		
			ORG 1990H		
1990	BD 30 9B	TABLE:	DEFB 0BDH,30H,9BH		
1993	BA 36 AE		DEFB 0BAH,36H,0AEH		
1996	AF 38 BF		DEFB 0AFH,39H,0BFH		
1999	BE		DEFB 0BEH		

Procedure

(a) Enter or download Program 2 into the Micro-Professor.

(b) Enter a two digit BCD number into address 1A00 hex.

(c) Now run the program. When it halts, press [MONI] and then examine the contents of addresses 1A10H and 1A11H. They should contain the 7-segment decoded versions of the numbers placed in address 1A00H. Try a few different numbers and check that they are all correct.

──────── **Question** ────────

6.12 How could the program be modified so that all the hexadecimal characters could be converted rather than just the numbers 0–9?

(d) Change the program into a subroutine by replacing the HALT instruction at the end with a RETURN. Notice that the subroutine will have to start at BCD1, address 1966H since the first two instructions have to be removed.

──────── **Questions** ────────

6.13 Write a **main** program starting at address 1800 hex which performs as follows:

(a) Converts the six digit BCD number in addresses 1A00–1A02H into six 7-segment decoded digits in addresses 1A10–1A15 hex.Use BCD1 as the conversion subroutine.

(b) Loads index register IX and 1A10 hex.

(c) Calls the SCAN1 routine at address 0624 hex.

(d) Jumps to step (a).

This program should put the numbers from the BCD buffer on to the Micro-Professor display.

6.14 Write a program which displays the contents of the accumulator in hexadecimal on the two right-hand 7-segment displays of the Micro-Professor.

Summary

Although all the data manipulation within a computer is performed in binary it is often necessary to use different types of code to make data processing easier.

Whenever a human interface is required, computers are generally expected to produce numbers in decimal form. This requires some means of conversion between binary and decimal. In particular, the decimal numbers may be conveniently stored in binary coded decimal (BCD) format. Each half of an 8-bit data byte can store the binary code equivalent to one decimal digit.

Conversion between binary and BCD may be performed by a shift and decimal adjust method.

Before these BCD numbers can be shown on a 7-segment display, they must be further converted into 7-segment code. This is the code that must be sent to an output port to illuminate the required character on the display. The conversion process simply requires a **look-up** table containing all of the 7-segment codes in the correct order.

SEVEN

Analogue and digital input/output

OBJECTIVES

When you have completed this chapter, you should be able to:

1. *Use software to control analogue to digital conversion hardware and read analogue input voltages.*
2. *Employ an analogue-to-digital converter to create a simple digital voltmeter.*
3. *Program a PIO device for various configurations of digital input and output.*
4. *Use digital signals of variable pulse width to control the speed of a small motor.*

EQUIPMENT REQUIRED

To complete all of the practical exercises in this chapter, you will need:

(a) *Micro-Professor MPF-1B.*
(b) *Applications board MAB.*
(c) *The POLSU replacement EPROM set.*

You may also wish to use a cross-assembler system, for which you will require a PC compatible computer and a Z80 assembler, editor and linker software together with a connecting lead.

The programs in this chapter are also available on disk.

PREREQUISITES

Before studying this chapter you should have:

(a) *An appreciation of the methods of analogue-to-digital and digital-to-analogue conversion.*
(b) *Some knowledge of the use of monitor subroutines.*

(c) *An understanding of the need to convert numbers between different formats.*
(d) *A basic knowledge of the use of programmable input/output devices.*

This could be achieved by studying Chapters 3 and 6 of the accompanying volume in the series, *Microelectronics NIII*.

7.1 INTRODUCTION

Most 'real world' signals are analogue in nature and cannot be directly processed by a computer until they are converted into their digital equivalent. This analogue-to-digital conversion process is vital to many parts of industry where signals must be measured and calculations performed on the basis of their value. Many processes depend upon the accuracy of the analogue-to-digital converters they employ.

This chapter shows how the computer can handle such conversions with the minimum of hardware and yet produce a reasonably accurate result. The analogue voltages can be simply measured, or they can be used to control the processors such as the speed of a motor. When analogue voltages are displayed, the computer system must perform a number of data conversion processes from Binary to BCD and to 7-Segment format, and these operations are examined.

Computers must be capable of handling both digital and analogue signals. It is important that digital input/output devices are made as flexible as possible to cater for varying requirements. Programmable input/output ports provide this flexibility, but they also require careful initialisation before they can achieve their correct operation. One such programmable device is examined in detail in this chapter.

PRACTICAL EXERCISE 7.1 – ANALOGUE-TO-DIGITAL CONVERSION (RAMP METHOD)

Experimental Concepts

When a computer is used to control an industrial process there is often a requirement to be able to input data from analogue devices. Since the computer can only process data in digital form an interface has to be provided which will perform the analogue-to-digital conversion. Integrated circuits exist that include the complete analogue interface but it is also possible to use the computer together with a digital-to-analogue converter and comparator to do the same thing.

In *Figure 7.1* digital outputs sent to port 81 hex are converted to their equivalent analogue voltage. This voltage is compared with the unknown analogue input in the comparator. If the unknown analogue input is **higher** than that generated by the computer the comparator output will be a logic '1'. If the unknown analogue input is **lower** than that generated by the computer then the comparator output will be a logic '0'. The computer can check the comparator output at any time since it is connected directly to bit 3 of port 80 hex.

- Computer output **too low** – logic '1' to bit 3 of port 80H
- Computer output **too high** – Logic '0' to bit 3 of port 80H

One conversion technique that is frequently employed is to output digital values to the output port starting with 00 and increasing by 1 on each step. After each value is output, the input from port 80H is checked to determine whether the output value is lower or higher than the unknown analogue input. As soon as the computer generates a voltage higher than the analogue input the process is stopped. This is shown by the waveforms in *Figure 7.2* (overleaf).

This method is described as the **ramp** method. The flow chart is shown in *Figure 7.3* (overleaf).

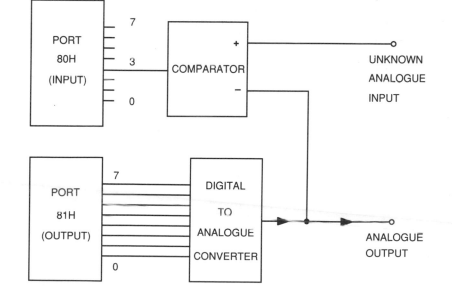

Figure 7.1 ADC hardware

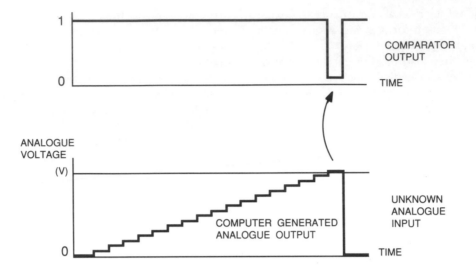

Figure 7.2   Ramp conversion waveforms

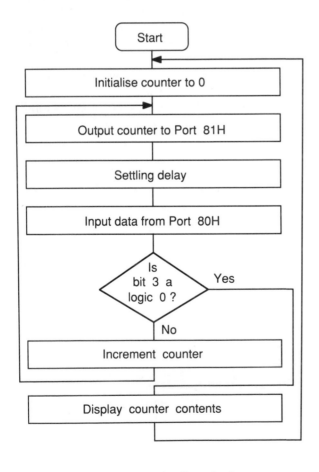

Figure 7.3   Ramp conversion flow chart

Program

The program below is based on *Figure 7.3*. Register C has been used as the counter. The display subroutine DISPA is the same one that was introduced in Practical Exercise 4.4. Here it is necessary to give some visible indication of the analogue voltage and the simplest way is to display its hexadecimal equivalent. In a subsequent exercise it is possible to convert this to a true voltage reading.

	Mnemonic	*Comment*
	ORG 1800H	
DISPA:	EQU 2420H	
ATOD:	LD C,0	; C is the counter
ATD:	LD A,C	
	OUT (81H),A	; Output counter
	NOP	; Settling delay
	NOP	
	IN A,(80H)	; Get Input from comparator
	AND 08H	; MASK BIT 3
	JP Z,DIS	; Jump if finished
	JP ATD	; Next value
DIS:	LD A,C	
	CALL	
	DISPA	; Display value
	JP ATOD	

Procedure

(a) Use the program mnemonics to assemble an equivalent machine code program starting at address 1800 hex.

(b) Enter or download the program into your Micro-Professor.

(c) Before executing the program, make sure your applications board is connected and that all the switches are in the OFF position, apart from:

ADC – ON

The analogue input selection switch should be set to position 3, i.e. the voltage input from the potentiometer.

(d) Now execute the program. The lights on the applications board should come on or flash and a hexadecimal number should appear in the two right-hand displays.

Turn the voltage control on the applications board and check that the number on the display changes.

———— Question ————

7.1 What is the range of values which can be obtained by turning the voltage control from one extreme to the other?

(e) The program operates relatively quickly but it can easily be slowed down to observe how the computer output rises to the unknown input then stops and start again.

Replace the three NOP instructions with a CALL instruction to a **delay** of about 10 or 20 ms.

(f) Run the program again and watch the lights of port 81H. When they reach the digital equivalent of the analogue input they go back to zero and start again.

Note that the display is too dim to see while the program is operating so slowly.

(g) Now observe the analogue voltage being generated by the computer by looking at the bar graph display.

Switch the BAR GRAPH ON. Turn the voltage control and observe the effect on the bar graph display.

Now answer the following questions.

———— Questions ————

7.2 What would happen on the display if the program was running at full speed but the analogue input voltage was higher than the voltage that the computer could produce?

7.3 If the program loop takes about 30 μs to execute once, what is the maximum time required for a complete conversion?

7.4 What is the range of digital values generated by the LIGHT sensor on the applications board when exposed to very bright and very dim light? Change the analogue input selector switch to position 4 to observe this.

7.5 What is the main drawback with this type of A to D software?

Summary

A digital-to-analogue converter and a comparator can be connected to a computer in such a way that an analogue-to-digital converter can be implemented. The hardware may be utilised in several ways to provide the digital equivalent of an unknown analogue input.

One conversion technique is known as the RAMP method in which the computer generates a steadily increasing voltage until it is found to be marginally greater than the input voltage being measured.

Analogue-to-digital converters are vital elements of many types of industrial equipment where process variables have to be measured and appropriate action taken.

PRACTICAL EXERCISE 7.2 – DIGITAL VOLTMETER

Experimental Concepts

This exercise is designed to show how the concepts explained in the previous practical exercise can be brought together with previous work in a practical application.

The exercise involves writing a program which makes the Micro-Professor and applications board act as a digital voltmeter (*Figure 7.4*). It will indicate the analogue voltage present at the input to the A–D converter on the applications board and display it as a number between 0.00 and 2.55 on the Micro-Professor display.

Previous exercises have involved the use of some of the system subroutines which will help to simplify the task here.

(a) Name: ANALOG
    Address: 2001 hex
    Function: Convert the analogue input into an 8-bit digital value.
    Input: None
    Output: Register A contains the digital equivalent of the analogue input between 00 and FF hex.
    Registers destroyed: AF
    Note: This subroutine uses a successive approximation algorithm which is significantly faster than the RAMP method of ADC.

(b) Name: SCAN1
    Address: 0624 hex
    Function: Scan keyboard and DISPLAY once from right to left. Execution time is 9.97 ms.

Input: IX points to the start of the display buffer.
Output: (a) If no key is pressed then carry flag = 1
           (b) If a key is pressed then carry flag = 0 and A contains the key code.
Registers destroyed: AF, AF', BC', DE'.
Note: (i) Six bytes are required for the display buffer to store the 7-segment patterns to be displayed.
           (ii) IX points to the right most digit IX + 5 points to the left most digit.

(c) Name: BINBCD
    Address: 2059 hex
    Function: Converts an 8 bit binary number into three digit BCD format between 000 and 255.
    Input: The binary number to be converted must be in address 1A00 hex.
    Output: The BCD number is stored in address 1A01 hex and the low half of address 1A02H with the Least Significant digit in bits 0–3 of address 1A01 hex.
    Registers destroyed: None

(d) Name: HEX7SG
    Address: 0678 hex
    Function: Converts a two digit hex or BCD number into 7-segment display format.
    Input: The first number is found in bits 0–3 of A.

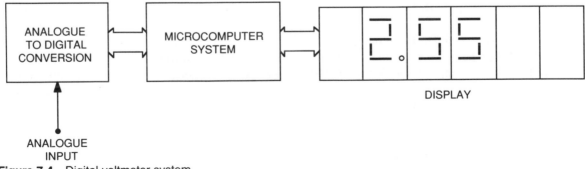

Figure 7.4 Digital voltmeter system

The second number is found in bits 4–7 of A.

HL must point to the required destination address of the first 7-segment decoded digit.

Output: The first decoded digit is placed in (HL) and the second digit is placed in (HL + 1). HL is incremented by two in the process.

Registers destroyed: AF,HL

(e) Name: HEX7
Address: 0689 hex
Function: Converts a single digit hex or BCD number into 7-segment format.
Input: The number is found in bits 0–3 of A.
Output: The result is also stored in A.
Registers destroyed: AF

The only real problem with this experiment is in keeping track of where the data is at each stage of the operation. This requires a clear understanding of the operation of each of the subroutines mentioned above, particularly their input and output conditions.

Three areas of memory are required to store the data at different stages. These are:

- BINARY BUFFER – Address 1A00 hex
- BCD BUFFER – Addresses 1A01 and 1A02 hex
- 7-SEGMENT DATA BUFFER – Addresses 1A10 to 1A15 hex (6 bytes)

The flow chart is shown in *Figure 7.5*.

Procedure

(a) Write a program starting at address 1800 hex which will make the system behave as a digital voltmeter. It should be based on the flow chart in *Figure 7.5*.

(b) Make sure the switches on the applications board are in the following positions:

ADC – ON
Analogue input – Position 3 – Voltage

All the other switches should be OFF.

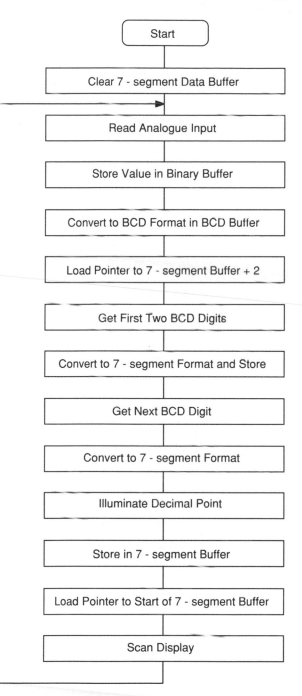

Figure 7.5 DVM flow chart

(c) Run the program. Turn the knob on the applications board and the number on the display should change accordingly.

If it does not go up to 2.55 don't worry. This

is simply due to the tolerance of the components on the board.

   Make sure that the digits are in the positions shown on the diagram in *Figure 7.4*.

(d) Change the position of the analogue input selector switch.

  (i) When connected to the LIGHT input, write down the range of voltages obtainable from the light transducer under a variety of light intensities.

  (ii) Write down the voltage generated by the temperature transducer when it is cold.

    Switch on the HEATER control switch and the heater should come on. If not make sure that the bit 5 switch of the port 80H is UP.

    Now write down the maximum voltage obtainable from the temperature transducer.

(e) It is possible to monitor external voltages between 0 and 2.55 volts with the system, but *great care must be taken* that high voltages are not applied.

   Switch the analogue input to the **external** position and connect pieces of wire to the analogue input and 0 volts screw terminals.

   Initially the system will read 2.55 volts because of an internal 'pull-up' resistor. Short the two wires together and a 0.00 volts reading should be obtained.

   Try to measure some low voltages. For example, try some 1.5 V batteries to find out their true voltage.

—— Question ——

7.6 If the analogue input to the ADC was divided by four, before it was measured, describe:

(a) The effect on the range of the DVM.

(b) The changes in the software which would have to take place.

Summary

An analogue-to-digital converter and microcomputer display can be readily used to produce a simple digital voltmeter as long as suitable software is provided. This software is required to perform a number of tasks including:

(a) Reading the analogue input and converting the data to binary form.

(b) Converting the binary data to BCD format.

(c) Converting the BCD data to 7-segment decoded format.

(d) Displaying the resulting numbers.

The system may be further expanded by providing suitable attenuators at the system input and using a multiplication routine within the program. This allows higher voltages to be monitored with some loss of accuracy.

PRACTICAL EXERCISE 7.3 – Z80 PIO PROGRAMMING

Experimental Concepts

The Z80 PIO is a relatively complex parallel input/output control chip which must be correctly programmed before it will function at all. Its internal structure is shown in *Figure 7.6*. Although it contains only two 8 bit data ports, it also contains two control ports which control the operation of the data ports. The data ports are **initialised** by sending certain bytes to the control ports.

   Generally, PIOs are arranged so that each of their four internal ports have separate consecutive addresses. The Micro-Professor is no exception and has the following port numbers allocated for the PIO:

Port A Data   – 80 hex
Port B Data   – 81 hex
Port A Control – 82 hex
Port B Control – 83 hex

Thus, to **initialise** port A, send the intialisation codes to port 82 hex and, to intialise port B, send them to port 83 hex. Each half of the PIO must be initialised separately.

   Since the two halves of the PIO are almost identical, only port A will be examined in further detail. The internal structure of port A is shown in *Figure 7.7* (page 88).

   The diagram shows both the control and data registers of port A. The data port is designated by the double line surrounding the registers. The function of each of the registers is given below.

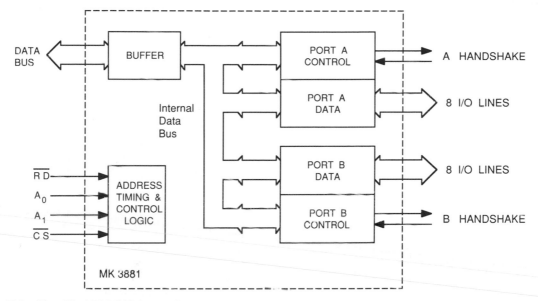

Figure 7.6 Simplified Z80 PIO internal structure

Data Port Registers

(a) *Output register* – This register latches output data from the CPU and sends it to any pin programmed to be an output.

(b) *Input register* – This register receives input data from external devices and may be read by the CPU at any time.

Control Registers

(A) Mode control register The way in which the PIO functions is determined by the data programmed into this two bit register. There are four possible operating modes (*Table 7.1*).

Modes 0 and 1 are the simplest operating modes and both require the operation of the PIO handshake lines for successful operation. Mode 2

Table 7.1 Operating modes of mode control register

Mode number	Mode register contents	Port operation
0	00	8 bits output
1	01	8 bits input
2	10	8 bits bidirectional
3	11	Bit control (inputs/outputs)

allows the PIO to operate both as an output and an input simultaneously so that it appears to be an extension of the system data bus.

Mode 3 operation is by far the most versatile. In this mode each individual bit can be programmed to be either an input or an output. Thus the 8-bit port may comprise, for example, six input bits and two output bits.

(B) Input/Output Select Register Whenever Mode 3 operation is programmed into the mode control register, the input/output select register must be programmed with a byte to determine which bits of the **data** port will be **inputs** and which bits will be **outputs**. A logic 1 indicates an input and a logic 0 indicates an output.

(C) Mask control register This register is only required when the PIO will be used to generate an **interrupt**. It contains two bits which have specific functions. One determines whether an interrupt will be generated if the input signal is at a logic 0 or 1 and the other determines whether AND logic or OR logic will be applied to the input signals to generate the interrupt.

(D) Mask register This register is programmed with a byte which determines which bits will be monitored for possible interrupt signals.

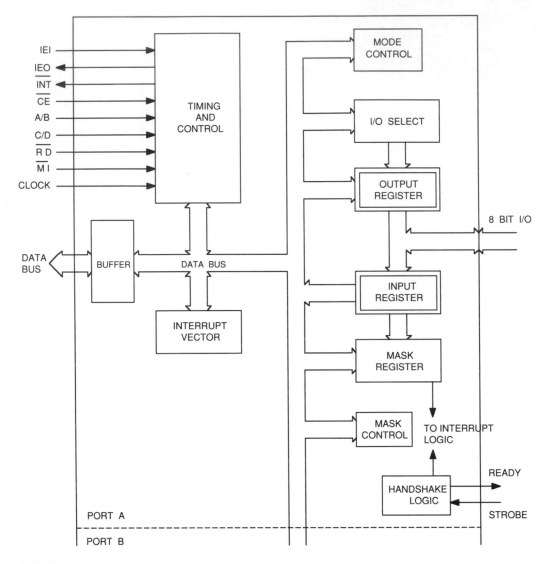

Figure 7.7   PIO port A structure

(E) Interrupt vector   Whenever interrupts are used, the interrupt vector register must be programmed with a number which will help to determine where an interrupt service subroutine will be found in memory.

PIO Programming

If the subject of interrupts is ignored, the following bytes of data must be sent to the control port for *each half* of the PIO.

(a) Send the mode control byte (*Figure 7.8*). The two most significant bits set the **mode** of the data port. Bits 5 and 4 can be anything and bits 0–3 must all be logic 1.

Figure 7.8   Z80 mode control byte

For example, the program:

LD A,0FH
OUT (83H),A

would set up port B of the PIO into the **output** mode.

Similarly,

LD A,0FFH
OUT (82H),A

would set up port A of the PIO into **bit control** mode, which is mode 3.

(b) When mode 3 is selected, the input/output select byte must be sent next (*Figure 7.9*). A logic 1 in any bit programs the corresponding bit of the data port to be an **input**. A logic 0 in any bit programs the corresponding bit of the data port to be an **output**.

Notice: 0 = Out; 1 = In.

I/O$_7$	I/O$_6$	I/O$_5$	I/O$_4$	I/O$_3$	I/O$_2$	I/O$_1$	I/O$_0$

Figure 7.9   Z80 I/O select byte

For example, to arrange bits 0–3 of port B as inputs and bits 4–7 to be outputs, send 0F hex:

LD A,0FH
OUT (83H),A

The following exercises involve programming the PIO in different ways and observing the effects produced. Remember that each time [RESET] is pressed, the PIO reverts to being 8 bits **input** on port A and 8 bits **output** on port B.

Procedure

(a) Write an initialisation program which makes bits 0–3 of port 81H all inputs and bits 4–7 of port 81H all outputs.

The program continues by sending the byte AA hex to port 81H and then halts.

(b) Convert the program to machine code, then download or enter it into the computer and run it.

───────── **Question** ─────────

7.7   What outputs are seen on the lights? Why?

────────────────────────────

(c) Change the initialisation program so that the function of each bit is reversed, then re-run the program. Does it perform as expected?

───────── **Question** ─────────

7.8   Describe what happens when a program is written to perform as follows:

(a) Initialise port 81H, bits 0–3 inputs, bits 4–7 outputs.
(b) Send byte AA hex to port.
(c) Delay for about one second.
(d) Initialise port 81H, bits 0–3 outputs, bits 4–7 inputs.
(e) Halt.

Note that the data byte is sent only once.

────────────────────────────

(d) Write an initialisation program such that bits 7, 6, and 5 of port 80H are arranged to be outputs and all other bits of port 80H are arranged to be inputs.

Continue the program by sending the byte FF hex to the port, then input the data from the port into the accumulator and halt.

Note: Although port 80H has a switch connected to each bit, it is quite safe to configure the bit as an **output**. This is because the special circuit on the applications board not only protects the PIO output but also allows it to override the switch data.

(e) Convert the program to machine code and download or enter it into the computer.

Before running the program, set the input switches to 07 hex.

(f) Now run the program, and when the HALT light comes on, press [MONI] to regain monitor control.

Examine the accumulator contents.

—————— **Questions** ——————

7.9 Explain why the accumulator contents are not the same as the data set on the input switches.

7.10 The **heater** on the applications board is connected to bit 5 of port 80H. A logic 1 on bit 5 turns the heater on and a logic 0 turns it OFF. Write a program which turns it ON if bits 0, 1 and 2 of port 80H are all at logic 1. If they are not *all* at logic 1 it must be turned OFF. The program then reads the data from the switches and sends it to port 81H. It should run continuously.

Remember to turn the HEATER control switch ON before running the program on the computer.

Summary

Programmable input/output ports have many applications in microprocessor based systems because of the flexibility which they allow. A simple microprocessor based controller may easily be configured by suitable software for a wide range of functions.

A typical PIO contains two data ports and two associated control ports.

Programming a PIO is accomplished by sending specific codes to the control port associated with each data port. These initially set the operating mode to one of the following:

(a) Output on 8 bits   – Mode 0
(b) Input on 8 bits   – Mode 1
(c) Bidirectional on 8 bits   – Mode 2
(d) Bit control   – Mode 3

When mode 3 is selected, the next byte sent to the control port determines which bits will be regarded as inputs and which will be regarded as outputs. A logic 1 produces an **input** bit and a logic 0 produces an **output** bit.

If a port is configured with some bits as outputs and some as inputs, whenever an **input** instruction is executed, the accumulator contents will contain the following: data from the input pins for all bits designated as inputs, plus the last data output to the bits designated as outputs.

PRACTICAL EXERCISE 7.4 – MOTOR SPEED CONTROL

Experimental Concepts

When the PIO on the Micro-Professor is configured so that bits 6 and 7 of port A are **outputs** the codes sent to those bits control the direction of the **motor** on the applications board. In the previous motor exercises all that has been required is to turn the motor on and off either after a certain time has elapsed or after a certain number of revolutions. However, if the signal to these bits is changed quite rapidly the **speed** of the motor may be controlled with reasonable accuracy.

For example, if the signal sent to bit 7 of port A is a squarewave, while bit 6 remains at a logic 0, the motor will be turned on and off very rapidly so that it will run at about half its normal full speed. If it is turned off for longer than it is turned on, it should run more slowly, and so on. The ratio of ON to OFF time will determine how fast the motor will run. This is illustrated in *Figure 7.10*.

The simplest speed control system uses a counter which counts continuously. It therefore goes from 0 to 255 and then immediately back to 0 again, and continues counting indefinitely. Each time the counter reaches 0 the motor control is turned ON, and each time it reaches a predetermined number it is turned OFF. Thus the higher the number, the longer the motor is ON and the higher the speed.

In this exercise the number will be obtained from the input switches connected to bits 0–3 of port 80H, but it could be obtained as the result of a calculation or, for example from an analogue input control.

Remember that the motor can be turned ON by sending code 80 hex to port 80H, and can be turned OFF by sending code 00 hex to port 80H (*Figure 7.11*).

Notice that the program runs continuously. If bits 0–3 of the input port are to be used to control

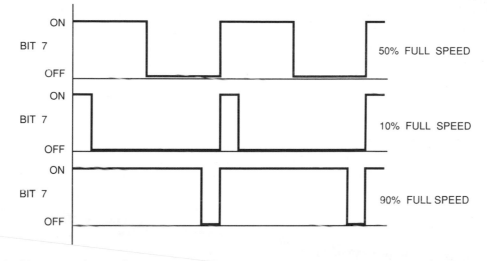

ON
BIT 7
OFF
50% FULL SPEED

ON
BIT 7
OFF
10% FULL SPEED

ON
BIT 7
OFF
90% FULL SPEED

Figure 7.10 Motor speed control

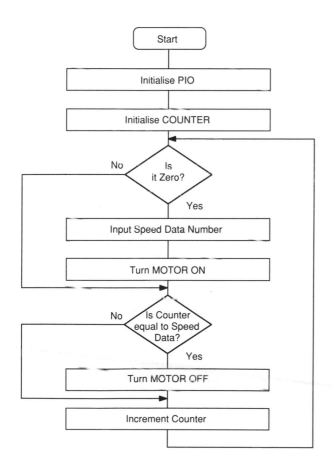

Figure 7.11 Flow chart

the motor speed then they must be rotated four places left before they are compared with the **counter**. This will allow a speed variation between 0 per cent and approximately 90 per cent of full speed (0/16 per cent and 15/16 per cent). A new number is input for the speed data only when the **counter** reaches 0. This will allow the main control loop to run as quickly as possible and may be important if the program is modified so that it takes more than a single instruction to get the input control data. Remember to use logical instructions to **mask** unwanted bits.

Procedure

(a) First work out the initialisation program which will be required for the PIO. Use mode 3.
(b) Now write an assembly language program based on the rest of *Figure 7.11*, and then assemble it starting at address 1800 hex.
(c) Download or enter the program into the Micro-Professor and execute it, making sure that the MOTOR control switch is ON.

Question

7.11 What is the lowest binary input on bits 0–3 of port 80H which makes the motor turn?

(d) Modify the program so that the data input on bit 5 of port 80H controls the *direction* of the motor rotation. This should be checked only when the COUNTER reaches zero:

BIT 5 = 1 should give FORWARD rotation
BIT 5 = 0 should give REVERSE rotation

Does the program perform as expected?

(e) Now modify the program again so that the speed control number is obtained from the **analogue** input. To do this simply use the ANALOG subroutine call by including in the program the instruction:

CALL 2001H

This subroutine returns a number between 0 and 255 in the accumulator depending upon the value of the analogue input.

Before executing the program make sure that the ADC control switch is ON and the analogue input selector slide switch is in position 3 – voltage. Turn the voltage control knob on the applications board and note how it affects the motor speed.

(f) Connect the **logic probe** lead to the socket on the applications board. Using this probe it is possible to see the effect of the program on the on/off ratio of the waveform controlling the motor.

Place the logic probe on the right-hand end of resistor R_{46} near the lower left-hand corner of the applications board. The **high** and **low** lights should be flashing if the motor is running. If not, try the probe on the right-hand end of resistor R_{47}.

Now turn the voltage control knob and observe the effect on the high and low lights of the logic probe as the speed changes. The relative brightness of the lights indicates how long the waveform spends at each logic level. At low speeds it should be mainly **low**, while at high speeds it should be mainly **high**.

——— Question ———

7.12 Why does the **motor** not run smoothly right down to very low speeds?

The following program illustrates the way in which the bits of the PIO ports can be changed from outputs to inputs and back to outputs again during the operation of a program.

——— Question ———

7.13 Write a program that controls the speed of the motor on the applications board according to the binary data set on bits 7, 6 and 5 of port 80H.

Note that the function of these bits will need to be changed during the program execution. Don't forget to change them back again!

Summary

The speed of a small motor can be controlled by sending a pulse waveform to it instead of a steady d.c. signal. The ratio of ON to OFF time of the pulse waveform determines the percentage of full speed at which the motor will run. Since the data direction of a PIO pin can be changed during program execution it is possible to use the same bit to control the motor that is used to input speed data during another part of the program.

EIGHT

Interrupt driven systems

OBJECTIVES

When you have completed this chapter, you should be able to:

1. Initialise a CPU and PIO for use with interrupts.
2. Use a software routine to count the number of times an interrupt occurs at an input.
3. Initialise a counter timer chip for use in an interrupt driven system.
4. Use a counter timer to generate random delays.
5. Design a simple reaction timing system.
6. Measure the speed of a small motor.

EQUIPMENT REQUIRED

To complete all of the practical exercises in this chapter, you will need:

(a) Micro-Professor MPF-1B with CTC chip fitted.
(b) Applications board MAB.
(c) The POLSU replacement EPROM set.

You may also wish to use a cross-assembler, for which you will require a PC compatible computer and a Z80 assembler, editor and linker software together with a connecting lead. The programs in this chapter are also available on disk.

PREREQUISITES

Before studying this chapter, you should have:

(a) A thorough knowledge of Z80 PIO programming including the use of interrupts.
(b) An awareness of the structure and modes of operation of the Z80, etc.

(c) An ability to initialise the Z80 CTC for all its modes of operation including those requiring interrupts.
(d) An awareness of the need for data conversions between binary, BCD and 7-segment data formats.

This could be achieved by studying as far as Chapter 6 in the accompanying volume in the series, *Microelectronics NIII*.

8.1 INTRODUCTION

When computers are connected to industrial or commercial systems, a lot of their time is spent in processing input and output data. Often computers are forced to wait for inputs to arrive from peripheral devices and this is very wasteful of computer time. Therefore, many systems allow the processor to continue with its work and only transfer data when a peripheral interrupts the CPU. In this way very little time is actually spent dealing with mundane data transfers.

Setting up a system to operate with interrupts from peripherals is not a trivial matter and this chapter deals with some of the issues involved. In particular, the methods of initialising all the devices in a system to operate with interrupts is investigated. Also, the special requirements of an interrupt service routine are covered in practical cases.

The exercises cover a range of interesting systems. Two use the small motor on the applications board to generate interrupts via its propeller which can simply be counted, or can be timed to measure its speed. This is very similar to the systems frequently employed in industrial settings where a certain action must take place either for a fixed number of occurrences or at a predetermined rate.

The other exercise shows how a simple reaction timer can be constructed by using a counter timer chip to produce a random delay and then time the reaction of the user to a stimulus. It can be made accurate to the nearest one hundredth of a second.

PRACTICAL EXERCISE 8.1 – INTERRUPT DRIVEN REVOLUTION COUNTER

Experimental Concepts

Many systems require a microprocessor to count events and to update a display with the current count continuously. One way to achieve this is simply to produce a program that displays the count value but allow the program to be interrupted whenever a new value is counted.

This idea is investigated in this experiment. The propeller of the motor is used to cut the infra-red beam to generate a high speed pulse input.

The program is relatively straightforward, but a suitable initialisation program must be written for the PIO to allow interrupts to take place. A monitor routine can be used to scan the display and show the data. Monitor routines used previously can also be used to perform the data conversions between binary, BCD and 7-segment formats.

The flow chart for the interrupt-driven system is shown in *Figure 8.1*. The buffers required for this program are:

1A00 Hex – Propeller edge counter
1A01–1A02 Hex – Revolution counter
1A03–1A05 Hex – BCD of revolutions
1A06–1A0B Hex – 7-segment decoded data

In addition, a service routine start address table is required. This can be placed at addresses 1A10 and 1A11 hex.

PIO Initialisation

The connections between the motor and the PIO port A are shown in *Figure 8.2*.

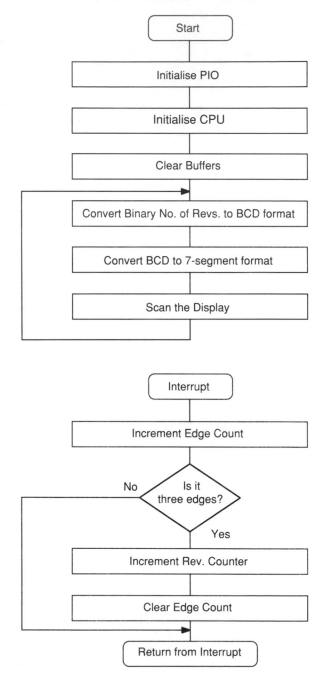

Figure 8.1   Revolution counting flow chart

The PIO must be initialised so that bits 6 and 7 are outputs and all the other bits are inputs. However, it should only allow a logic 0 on bit 4 to cause an interrupt. Notice how this is achieved in the initialisation part of the program.

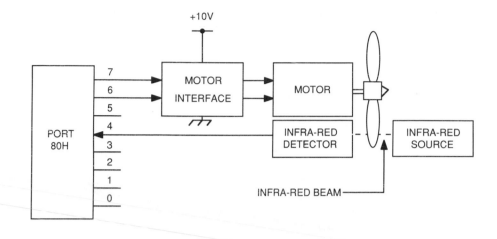

Figure 8.2   Motor and PIO connections

Program

	Mnemonic	Comment
	ORG 1800H	
HEX7SG:	EQU 0678H	
HEX7:	EQU 0689H	
SCAN1:	EQU 0624H	
;		
PIOINI:	LD A,0FFH	
	OUT (82H),A	; PORT 80H MODE 3
	LD A,3FH	
	OUT (82H),A	; BITS 6 AND 7 OUTPUTS
	LD A,10H	
	OUT (82H),A	; INTERRUPT VECTOR (1A) 10H
	LD A,97H	; IE, OR, LOW, MASK FOLLOWS
	OUT (82H),A	; INTERRUPT CONTROL
	LD A,0EFH	; 1 1 1 0 1 1 1 1
	OUT (82H),A	; INTERRUPT MASK BIT 4
;		
CPUINI:	LD A,1AH	
	LD I,A	; INTERRUPT REGISTER 1A
	LD HL,1890H	; 1890H SERVICE ROUTINE START
	LD (1A10H),HL	; LOAD START ADD. TABLE
	IM 2	; MODE 2
	EI	; ENABLE INTERRUPTS
;		
CLBUFS:	LD B,0DH	; BYTE COUNTER
	XOR A	; CLEAR A
	LD HL,1A00H	; BUFFER START
CLB:	LD (HL),A	; CLEAR MEMORY
	INC HL	
	DEC B	
	JP NZ,CLB	; LOOP UNTIL DONE

	Mnemonic	Comment
MSTART:	LD A,80H	; START MOTOR
	OUT (80H),A	;
;		
REPEAT	CALL BINCON	; BINARY TO BCD CONVERSION
;		
BCD7SG:	LD HL,1A06H	; DESTINATION POINTER
	LD A,(1A03H)	
	CALL HEX7SG	; CONVERT 2 RH DIGITS
	LD A,(1A04H)	
	CALL HEX7SG	; CONVERT MIDDLE DIGITS
	LD A,(1A05H)	
	CALL HEX7	
	LD (HL),A	
;		
SCAN:	LD IX,1A06H	
	CALL SCAN1	; SCAN1 ADDRESS 0624H
	JP REPEAT	
;		
BINCON:	DI	
	LD B,03	
	LD HL,1A03H	
BCL:	LD (HL),0	
	INC HL	
	DEC B	
	JP NZ,BCL	
	LD HL,(1A01H)	; 16 BITS INTO HL
	PUSH HL	
	LD C,10H	
LOOP:	LD HL,1A01H	; BINARY START ADD
	RL (HL)	
	INC HL	
	RL (HL)	
	INC HL	; HL HOLDS 1A03H
	LD B,03	; 3 BYTES OF BCD
BCDADJ:	LD A,(HL)	
	ADC A,A	
	DAA	
	LD (HL),A	
	INC HL	
	BJNZ BCDADJ	
	DEC C	
	JR NZ,LOOP	
	POP HL	
	LD (1A01H),HL	; RESTORE BINARY DATA
	EI	
	RET	
;		
; INTERRUPT SERVICE ROUTINE STARTS		
; AT ADDRESS 1890H		
	ORG 1890H	
;		

	Mnemonic	Comment
ISR:	PUSH AF	
	PUSH HL	
	LD HL,1A00H	; POINT TO EDGE COUNT
	INC (HL)	; INCREMENT IT
	LD A,(HL)	
	CP 03	; IS IT 3?
	JP NZ,NOT3	
	LD HL,(1A01H)	; GET REV COUNT
	INC HL	; INCREMENT IT
	LD (1A01H),HL	; RESTORE REV COUNT
	XOR A	
	LD (1A00H),A	; CLEAR EDGE COUNT
NOT3:	POP HL	
	POP AF	
	EI	
	RETI	; RETURN FROM INTERRUPT

Procedure

(a) After examining each part of the program in detail, assemble it then download or enter it into the Micro-Professor starting at address 1800 hex. Make sure that the interrupt service routine begins at address 1890 hex.

(b) Put the motor control switch down:

MOTOR – ON

(c) Run the program. The number on the display should be increasing rapidly as the propeller revolves. Slow it down or stop it with a finger to verify that one revolution is being counted for every three propeller blades which cut the infra-red beam.

(d) Care must be taken not to stop the program in the middle of the interrupt service routine, otherwise further interrupts will be locked out. Switch the motor off by the motor control switch and then, when it is stationary, press [MONI] to stop the program.

Work out the speed of the motor by performing a simple check on the number of revolutions counted in one minute.

Questions

8.1 Measure the speed of the motor a number of times while it is going 'forward', then modify the program so that the motor goes in 're-verse'. Does it go faster forwards or in reverse?

8.2 Why is it necessary to disable interrupts during the binary to BCD conversion part of the program?

Now try the following question based on the revolution counting program.

Question

8.3 Since the interrupt driven counting routine is independent of the motor control signals it is possible to monitor precisely the number of revolutions turned by the propeller, even after the 'stop' command has been sent to it. Modify the previous program so that a command is given to stop the motor after 100 revolutions have been counted. Note how many extra revolutions are measured.

Does the motor 'run on' by the same number of revolutions no matter what the number requested is? Try some higher and lower numbers.

Summary

One practical application of the use of an interrupt is to count the number of revolutions turned by a small d.c. motor. Each time a blade of the propeller cuts the infra-red beam, an interrupt signal is sent to the processor. If the processor is not performing a routine which uses the current count number, the 'blade edge' count is increased by one, and the revolution count is increased by one for every three blades which cut the beam.

The main advantage of using this technique compared with alternative methods is that the processor can continue with the task of multiplexing the display unless an interrupt occurs. This simplifies the software although the program is no shorter because of the addition of the initialisation routines for the PIO and CPU.

PRACTICAL EXERCISE 8.2 – REACTION TIMER

Experimental Concepts

In this exercise the Micro-Professor can be turned into a simple reaction timer. The Z80 CTC (counter timer) is used to generate known timing intervals of 0.01 seconds. These are used as the main timing reference in the system. This operates according to the flow chart in *Figure 8.3*.

Two timing functions are required in the program. One delays for a random time and the other measures the reaction time. In each case it is possible to write an interrupt service routine which simply keeps track of the number of 0.01 second intervals which have elapsed. The easiest way to do this is by decrementing a register pair which contains the current count. The CTC operating mode and the interrupt response must be set up first.

CTC programming Two control bytes must be sent to the counter timer channel to initialise it. Each channel in use requires initialisation, but this can be ignored for any channel which will not be employed in a particular program. In addition, the CTC must be sent an interrupt vector. This is sent to channel 0, but applies to the whole device.

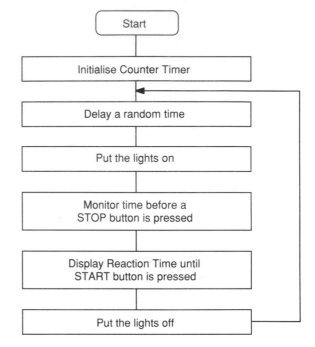

Figure 8.3 Reaction timer flow chart

Interrupt vector The CTC interrupt vector works in exactly the same way as that in the PIO which has been described already. It contains the low part of the address at which the address of the interrupt service routine will be found.

Since one interrupt vector is loaded for the complete CTC, some means of identifying an individual channel must be included. This is inserted automatically by the CTC as shown in *Figure 8.4*.

Thus any interrupt vector must be a multiple of 08 hex.

Note that in the Micro-Professor, the CTC channels 0–3 have addresses 40, 41, 42 and 43 hex. The interrupt vector must be sent to port 40 hex.

Channel control register The channel control register (*Figure 8.5*) sets up all the functions of the channel. Each bit of the byte sent to it is significant.

Bit 7   1 – Interrupt from this channel is enabled when the count is decremented to zero.

      0 – Interrupt from this channel is disabled.

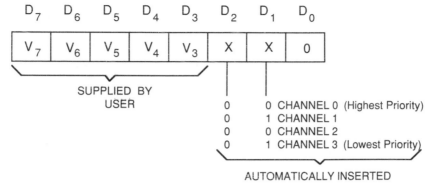

Figure 8.4 Interrupt vector register

Bit 6 1 – Counter mode selected. External CLK/TRG input decrements the counter.

0 – Timer mode is selected. The system clock decrements the counter via the prescaler. The ZC/TO pin has a pulse train output whose period is:

$$t \times p \times TC$$

where t = system clock period,
p = prescaler value (16 or 256),
TC = **time constant** value.

Bit 5 Defined in **timer mode** only. Ignored in **counter** mode.
1 – Prescaler set to 256
0 – Prescaler set to 16

Bit 4 1 – Positive edge trigger starts **timer** or decrements **counter**
0 – Negative edge trigger starts **timer** or decrements **counter**.

Bit 3 Defined for **timer mode** only. Ignored in **counter** mode.
1 – External trigger on CLK/TRG input starts timing.

0 – Timing starts in the machine cycle immediately after the **time constant** is loaded.

Bit 2 1 – **Time constant** byte is the next data to be written to the channel.
0 – No time constant follows. This is used when the control byte must be modified without affecting the time constant.

Bit 1 1 – **Reset** channel – the channel **stops** operation temporarily. If both Bit 2 and Bit 1 = 1, operation resumes when a new time constant is loaded.
0 – Channel continues current operation.

Bit 0 1 – It must always be a logic 1.

The descriptions may appear complex, but are not very difficult to put into practice.

For example, imagine that a **timer** is required which will start as soon as the time constant is loaded. Its prescaler value needs to be 16 since the time intervals required are very short, and no interrupts are required. The required control byte

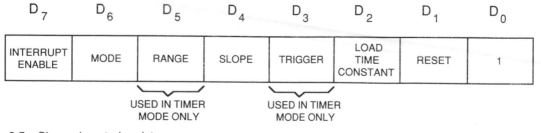

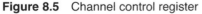

Figure 8.5 Channel control register

0	0	0	0	0	1	1	1	= 07 Hex
DI	TIMER	16	– VE TRIG	INT.	LOAD TC	RESET		

Figure 8.6 Control byte

would be as in *Figure 8.6* and this would be sent to the required channel, followed by the time constant byte.

Time constant register If bit 2 of the channel control byte is set to logic 1, then the next byte sent to the port must be the **time constant**. This is loaded with a number between 1 and 256.

Time constant	Decimal counts to zero
01	1
02	2
FF	255
00	256

Reaction timer In the reaction timer, the basic timing interval is 0.01 seconds. Therefore, the CTC must be programmed in **timer** mode. Since the CPU clock is 1.79 MHz, the value required for the time constant can be calculated as shown below. A prescaler value of 256 is needed for this length of time.

$$\frac{1}{1.79 \times 10^6} \times 256 \times TC = 0.01$$

Thus

$$TC = \frac{0.01 \times 1.79 \times 10^6}{256}$$

$$\underline{\underline{TC = 70}}$$

If an interrupt is generated every 0.01 seconds, the interrupt service routine can be made to decrement a count each time it runs. This count can be checked in the main program.

A **random** number may be obtained in the Z80 by loading the accumulator from the **refresh** register R. This is a register that holds a number between 0 and 127 which changes during every instruction execution. Therefore, a random time after the computer is switched on, it will contain a random number.

Study the program below and identify the various parts of it. It uses the following memory locations to store data:

1A01H–1A02H – **binary** number from HL for display

1A03H–1A05H – **BCD** version of binary in 1A01H–1A02H

1A06H–1A0AH – **7-segment** decoded BCD for display

Program

	Mnemonic	Comment
	ORG 1800H	
HEX7SG:	EQU 0678H	
HEX7:	EQU 0689H	
SCAN1:	EQU 0624H	
;		
CPUINI:	LD A,19H	
	LD I,A	; HIGH BYTE IN I
	LD HL,18B0H	

	Mnemonic	Comment
	LD (1960H),HL	; 1960H IS START ADDRESS
	IM2	; TABLE
	EI	
;		
CTCINI:	LD A,60H	; VECTOR
	OUT (40H),A	
	LD A,0B7H	; CONTROL BYTE
	OUT (40H),A	
	LD A,70	; TIME CONSTANT
	OUT (40H),A	
;		
RWAIT:	LD A,R	; GET RANDOM NUMBER
	LD L,A	
	LD H,0	
	ADD HL,HL	; × 2
	ADD HL,HL	; × 4
	ADD HL,HL	; × 8
	LD A,3FH	;
	OR L	
	LD L,A	; ENSURE 1 SEC MIN DELAY
;		
WLOOP:	PUSH HL	; TRANSFER FROM HL TO
	POP DE	; DE IN CASE INTERRUPT
	LD A,D	; CHANGES HL DURING TESTING
	OR E	; HL IS DECREMENTED
	JP NZ,WLOOP	; DURING INTERRUPTS
;		
LEDSON:	LD A,0FFH	; LIGHTS ON
	OUT (81H),A	
;		
TIME	LD HL,0	; WAIT UNTIL USER
TLOOP:	IN A,(00)	; KEY IS PRESSED
	AND 40H	
	JP NZ,TLOOP	
	LD A,0B3H	; RESET CTC
	OUT (40H),A	
;		
	XOR A	
INVHL:	EX DE,HL	; INVERT HL
	LD HL,0	
	SBC HL,DE	
;		
DIS:	CALL HLDISP	; DISPLAY HL
	JP C,DIS	; UNTIL GO KEY
	CP 16H	; PRESSED
	JP NZ,DIS	
	LD A,00	
	OUT (81H),A	
	JP CTCINI	
HLDISP:	LD (1A01H),HL	
	PUSH HL	; PRESERVE BINARY NO.

	Mnemonic	Comment
	LD B,03	
	LD HL,1A03H	
BCLEAR:	LD (HL),0	; CLEAR BCD DATA
	INC HL	
	DEC B	
	JP NZ,BCLEAR	
	LD C,10H	; 16 BITS
COLOOP:	LD HL,1A01H	; CONVERT DATA FROM
	RL (HL)	; BINARY IN 1A01H–1A02H
	INC HL	; TO BCD IN 1A03H–1A05H
	RL (HL)	
	INC HL	
	LD B,03	
BADJ:	LD A,(HL)	
	ADC A,A	
	DAA	
	LD (HL),A	
	INC HL	
	DEC B	
	JP NZ,BADJ	
	DEC C	
	JP NZ,COLOOP	
;		
BCD7SE	LD HL,1A06H	; DESTINATION POINTER
	LD A,(1A03H)	; GET 1ST BCD BYTE
	CALL HEX7SG	; HEX7SG = 0678H
	LD A,(1A04H)	; NEXT BCD BYTE
	CALL HEX7SG	
	LD A,(1A05H)	; LAST BCD BYTE
	CALL HEX7	; HEX7 = 0689H
	LD (1A0AH),A	
;		
	LD HL,1A08H	
	LD A,(HL)	
	OR 40H	; ILLUMINATE DECIMAL POINT
	LD (HL),A	
;		
SCAN:	LD IX,1A06H	; IX POINTS TO SCAN BUFFER
	CALL SCAN1	; SCAN1 = 0624H
	POP HL	
	LD (1A01H),HL	; RESTORE ORIG. BINARY NO.
	RET	
;		
;	AT ADDRESS 18B0H	
	ORG 18B0H	
ISR:	DEC HL	
	EI	
	RETI	

Procedure

(a) Analyse each part of the program. In particular make sure that the following points are understood:

  (i) How the control byte for the CTC is made up.

  (ii) Why HL is added to itself three times.

  (iii) Why HL is transferred into DE before it is tested.

(b) Now assemble the program and enter or download it into the Micro-Professor starting at address 1800 hex. Ensure that the interrupt service routine start address table starts at address 1960 hex and that it contains the correct start address of the service routine, wherever that happens to be located.

(c) Execute the program and try it out! The idea is that as soon as the lights on the applications board come on, the [USER] key must be pressed. The **reaction time** then shows on the display. To repeat the test simply press the [GO] key.

  Does it generate a sensible reaction time on the display?

  Check the output from the CTC channel 0 ZC/TO pin with the logic probe.

  Testing the family or friends can produce some very amusing results.

(d) Now answer the following questions.

———— Questions ————

8.4 What is the longest 'reaction time' this program can accommodate? Why?

8.5 Apart from the change required in the display format, how could the program be modified so that it was accurate to within 0.001 seconds?

Now try the following question. It should not involve major re-writing if it is planned carefully.

———— Question ————

8.6 How could the program be changed so that it becomes a simple stop/start timer? When the [GO] key is pressed the timer starts, and when the [USER] key is pressed again the timer stops.

PRACTICAL EXERCISE 8.3 – SPEED MEASUREMENT

Experimental Concepts

Measurement of the speed of a small motor appears to be a relatively simple concept. It requires some means of counting the number of revolutions in a fixed time. Fortunately, the CTC is ideal for this task since its two major functions are counting and timing.

The last exercise introduced the concept of timing by using the CTC to generate 0.01 second pulses. By counting 100 of these an accurate 1 second time interval can be generated. The speed of the motor revolutions per second could be calculated by counting the number of revolutions that occur in this time.

However, the propeller on the applications board has three blades. Therefore, it is only necessary to count the number of pulses generated in one third of a second to measure the speed in revolutions per second.

Channel 0 of the CTC is connected directly to the pulse input from the propeller blade infra-red detector. This channel can, therefore, be used directly as the counter (*Figure 8.7* overleaf).

Channel 1 of the CTC can be used as the timer and programmed to generate a timing interval of 0.01 seconds. When 33 such intervals have been counted, the number of pulses counted by Channel 0 can be displayed on the Micro-Professor display.

Calculations   The maximum speed of the motor should be about 12 000 r.p.m. This gives about 200 revolutions per second.

If the Counter, Channel 0, starts with a count of

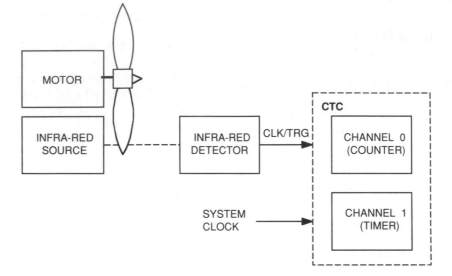

Figure 8.7   Speed measurement system

256 (actually 0 is programmed into the time constant register) then it will certainly not be decremented to zero in one third of a second. There is no need to bother with an interrupt service routine for this channel. It can simply be allowed to decrement the count for every pulse, and whenever the timer registers one third of a second it can be read and the speed calculated. The interrupt service routine for the timer, channel 0 has to stop the counter channel and then convert the number in the counter to a suitable form for the display.

This is shown in *Figure 8.8*.

Notes

(a) When the counter is operating, it counts **down** from 256. Therefore, to obtain the correct value for the binary count, the two's complement of the number must be obtained. Since only an 8-bit number is involved, the instructions NEG (hex code ED 44) may be used to achieve this.

(b) The PIO must be initialised for motor operation. This may be achieved by a **call** to address 2091 hex.

(c) Three buffers are required for binary, BCD and 7-segment decoded data. These should be:

1A00H   – Binary rev count
1A01H–1A02H – BCD Version of binary (000–255)
1900H–1905H – 6 digit 7-segment data

(d) The subroutine at address 2059H may be used to convert binary data in address 1A00H to BCD data in addresses 1A01–1A02H as detailed below.

Name: BINBCD
Address: 2059H
Function: 8 bit binary to BCD conversion
Input: Binary number in address 1A00H
Output: Packed BCD number in addresses 1A01H and 1A02H. The least significant digit is returned to bits 0–3 of address 1A01H, the next digit in bits 4–7 of address 1A01H and the most significant digit in bits 0–3 of address 1A02H.
Registers affected: None.

Note, however, that the binary data in address 1A00H is destroyed during the subroutine. It should, therefore, be preserved on the stack before the routine and restored afterwards.

(e) The subroutine at address 207DH can be used to clear all of the above buffers.

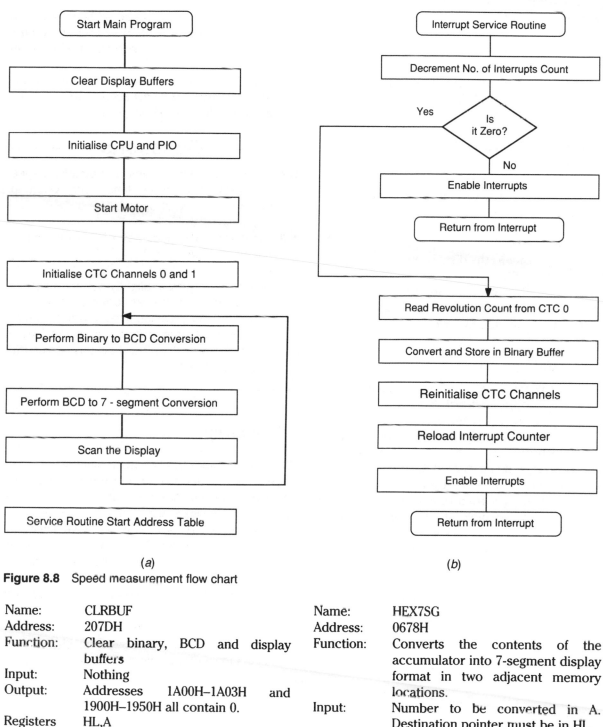

(a)   *(b)*

Figure 8.8 Speed measurement flow chart

Name: CLRBUF
Address: 207DH
Function: Clear binary, BCD and display buffers
Input: Nothing
Output: Addresses 1A00H–1A03H and 1900H–1950H all contain 0.
Registers affected: HL,A

(f) The hexadecimal to 7-segment subroutines may be used to simplify the display conversion routines.

Name: HEX7SG
Address: 0678H
Function: Converts the contents of the accumulator into 7-segment display format in two adjacent memory locations.
Input: Number to be converted in A. Destination pointer must be in HL.
Output: Least significant number is 7-segment format in (HL) and the next most significant digit in (HL + 1). HL is incremented twice.

Registers affected: AF,HL

Name: HEX7

Address: 0689H

Function: Converts the hex or BCD number in bits of A into 7-segment decoded format.

Input: Number in bits 0–3 of A.

Output: 7-segment decoded data in A.

Registers affected: AF

(g) The display scanning may be simplified by using the SCAN1 subroutine.

Name: SCAN1

Address: 0624H

Function: Scans the display once from right to left.

Input: IX points to scan buffer.

Output: None relevant to this application.

Registers affected: AF, AF', BC', DE'

Procedure

(a) With all the clues given in the notes on the flow chart the first task is to write a suitable program to display the motor speed.

Proceed carefully with the program using the subroutines whenever possible to simplify its operation. Be careful to note that some routines may affect registers!

Write the program in mnemonic form first and then assemble it starting at address 1800H.

(b) Enter or download the program into the Micro-Professor and try it out.

If it runs first time, then congratulations are in order! If not, check it through carefully and if all else fails, look in the answers at the back of the book.

(c) The number on the display may not be very steady since the motor speed fluctuates con-siderably. However, it should be possible to slow it down with light finger pressure on the propeller and see a significant change in value.

Write down the maximum speed of the motor in revolutions per second and work out the speed in revolutions per minute.

———— Questions ————

8.7 Discuss the advantages and disadvantages of choosing a longer time over which to measure the motor speed.

8.8 What limits the accuracy of this method of speed measurement?

8.9 Modify the program slightly so that it counts the number of pulses generated over a period of 0.66 seconds. Then divide the number by two before it is stored in the binary buffer. Comment on the effect this has on the displayed numbers.

Summary

The counter timer has two main functions, counting external events and timing intervals. These are precisely the functions required in a system designed to measure the speed of a motor. One channel of the CTC is used to measure a precise time interval of 0.33 seconds and another is used to count the number of pulses generated by the 3-bladed propeller during this time. The number of pulses is equal to the speed in revolutions per second of the motor.

When the speed is measured its value can be converted from binary to 7-segment display format and shown on the Micro-Professor display.

Appendix

OPTIONAL EXERCISE – USING A CROSS-ASSEMBLER

Experimental Concepts

The exercises in this book have been written so that they can be used with or without the benefit of a cross-assembler. Maximum benefit can be obtained from the exercises if a cross-assembler is used because it removes the problem of having to look up hexadecimal codes and it also allows the programs to be stored on a host computer disk system.

This exercise introduces the techniques and equipment required to use a **cross-assembler** package. It covers both the **physical** connections of the computer to the Micro-Professor and the **software** required.

A relatively trivial example is used for the program so that the software does not complicate the method.

The object of this exercise is simply to write the basic input/output program below, transfer it to the Micro-Professor and to execute it.

Program

```
PE1;    ORG 1800H
;
START: IN A,(80H)   ; GET INPUT PORT DATA
       XOR A,0FH    ; INVERT HALF
       OUT (81H),A  ; OUTPUT IT
       JP START     ; REPEAT INDEFINITELY
```

Procedure

Follow through this procedure carefully. The first steps need only to be performed once although the rest will apply whenever a program has to be downloaded:

(a) The first step is to ensure that the host computer and the Micro-Professor are connected together via a suitable cable:

    (i) *IBM PC or compatible*. If the host computer is an IBM PC or compatible, it is necessary to connect either COM1 or COM2 to the Micro-Professor EAR socket. The necessary pin connections are shown in *Figure A1.1*.

    (ii) *BBC computer*. If a BBC computer is used as the host, the arrangement is similar except that the pin configuration is different (*Figure A1.2*). Suitable cables are available from Flight Electronics Ltd. Connect the cable between the host PC socket and the Micro-Professor.

(b) The next step is to ensure that the **baud rate** of the serial connection circuits is correctly initialised.

The Micro-Professor expects to receive data at 2400 baud. This means that data must be transmitted at the same rate since there are no facilities for 'handshaking', i.e. regulating the flow of data.

In an IBM PC this is achieved by executing the command:

$$MODE \quad COM1:24,n,8,1$$

Ensure that the MODE.EXE file is present on disk before this command is executed. If the same PC is to be used for writing all the Z80 programs it would probably be worthwhile including this command in the

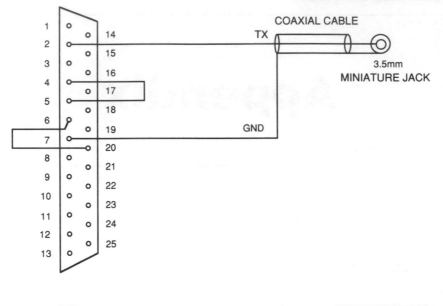

D25
IBM PC

MICRO-PROFESSOR
'EAR' SOCKET

Figure A1.1   IBM PC to Micro-Professor cable

AUTOEXEC.BAT file, or writing a short batch file to include it.

With a BBC computer, the relevant commands which initialise the serial port are:

*FX 5,2
*FX 8,5

(c) The next step in the process is to use a **word-**

processor or **text editor** to write the required program. Here, there are many possibilities.

IBM PC or compatible users may use a popular program such as Wordstar, Wordperfect, Sidekick, etc. or any program which is capable of producing an ASCII text file. Even Edlin will do!

RS423

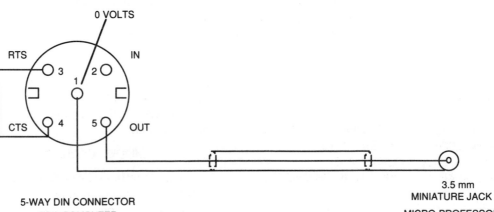

5-WAY DIN CONNECTOR
BBC COMPUTER

3.5 mm
MINIATURE JACK

MICRO-PROFESSOR
'EAR' SOCKET

Figure A1.2   BBC to Micro-Professor cable

BBC users may choose View or Wordwise or the BASIC text editing facilities if the Flight Z80 cross assembler package is being used.

With the chosen wordprocessor, enter the text of the program exactly as it is written, leaving appropriate spaces, and adding semi-colons as indicated before comments. Be careful to ensure that the assembly language syntax follows the required layout rules.

Before leaving the wordprocessor, ensure that the text is saved on disk in an appropriately named file. Careful use of the wordprocessor can help to minimise text entry by allowing blocks of text to be copied from one file to another, where sections are repeated. Remember that many IBM assemblers require a .ASM file extension. They are also likely to require text in ASCII format.

(d) Next, the wordprocessed text file in ASCII format must be assembled to create the appropriate .HEX, .OBJ and .PRN files.

Each assembler may be slightly different in format, so it is important to read the manuals supplied with the software.

For example, with the 2500 AD software assembler used by the author on an IBM compatible, the appropriate command would be:

C > X80

This command starts the assembler, which then prompts the user for input and output filenames and the required options.

When it has completed its operation, the assembler leaves a number of extra files on disk. If the original program was PROG1.ASM, the assembler would create:

PROG1.OBJ
and   PROG1.LST

Some assemblers also create a symbol cross-reference file known as

PROG1.XRF

(e) The next step is to LINK the .OBJ file to create a .HEX file. This requires the LINK.EXE program which comes with the operating system. The appropriate command would be:

C > LINK PROG1

The Linker program takes PROG1.OBJ and creates an Intel standard Hex file which can be downloaded into the Micro-Professor.

If the complete program is made up of various modules which have been created separately, the Link program can link them into a suitable HEX file ready for downloading.

With the Flight cross-assembler on a BBC computer, the procedure for writing and assembling programs is rather different. For precise instructions refer to the manual supplied with the system. This also applies to the downloading process which follows, although here the method is very similar whatever machine is used as the host.

(f) *The downloading process.* When the program has been assembled with no errors and the **linker** or the BBC cross-assembler has created the appropriate HEX file, the downloading to the Micro-Professor RAM can begin.

There are two stages – first prepare the Micro-Professor, then send the file.

First, **execute** the receive program on the Micro-Professor, by entering the command:

[ADDR] [2] [8] [0] [0] [GO]

The display will respond with the message:

OFFSEt

If no OFFSEt is required, press [GO], otherwise type in a 4-digit hexadecimal number. The program would then load at the address ORIGIN + OFFSET.

Note that if the ORIGIN has been set in the assembly language file at 1800H, then there is no need to add an offset.

The display will then show

SOUnd

To request a sound confirmation of the data transfer, press [GO], otherwise press any other key.

The display will then show SEnd for about 1 second followed by . . . .

This means that the system is waiting for a HEX file to be transmitted.

Now use an appropriate command on the

host system to send the file. For example, on an IBM PC or compatible, this would be:

C > COPY PROG1.HEX COM1:

When the **return** key is pressed, the file is transmitted to the Micro-Professor, and if it is received successfully, the display will show:

Good

for about one second. The Micro-Professor can then be used as normal, as though the program had been entered into RAM by hand.

A bad data transfer may result in an error message such as:

RECORD   – indicating that the INTEX HEX
                  format was incorrect;
or CHECSU – indicating a checksum error.

In either case, some data may have been transferred correctly, but it would be advisable to attempt the whole transfer again. After displaying an error message, the Micro-Professor returns to the download state to allow another attempt to be made.

(g) When downloading is complete, examine the program by stepping through the Micro-Professor memory starting at 1800H, and compare the data with what was expected from the program. This is normally done by **typing** the PROG1.LST file on the computer.

(h) Run the program and check whether it performs as expected.

(i) If all the operations have worked correctly so far, then try some modifications to the program and go through the whole process a few times until it is very familiar.

Modifications can either be made at assembly language level, in which case they would be stored on disk, or they can be made directly in the Micro-Professor memory. This, however, is only recommended for very minor changes or direct data entry.

Summary

Programs can be written in assembly language on a host machine, assembled, linked and then downloaded to a target system.

The process steps are to:

(a) Write the program in assembly language with a **wordprocessor**.
(b) Assemble the program with an **assember**.
(c) Link the program segment(s) with a **linker**.
(d) Download the program by (i) preparing the target system to receive data, then (ii) sending the file.
(e) Run the program using the Micro-Professor **monitor**.

Program changes can be made either by changing the assembly language file and repeating the whole process, or for minor changes, the **monitor** may be used to modify the machine code directly.

Answers

CHAPTER 1

1.1 When the switch is down the light now comes on instead of being off.

1.2 (a) 0 0 0 0 1 1 1 1.   (b) 1 0 1 0 1 0 1 0.
(c) 1 1 1 1 1 1 1 1.
If you have difficulty modifying the program in Part 3 press the following keys:

> [RESET]
> [ADDR]
> [0] [A] [8] [0]
> [GO]

1.3 Switch 0 affects light 7. Thereafter, switch 1 affects light 0, switch 2 affects light 1, etc.

1.4 (a) 1 0 0 0 0 0 0 0.   (b) 0 1 0 0 0 0 0 0.
If you have difficulty in modifying the program in Part 5, press the following keys:

> [RESET]
> [ADDR]
> [0] [A] [9] [8]
> [G0]

1.5 Lights 1, 3, 5 and 7 are ON all the time. Lights 0, 2, 4 and 6 can be controlled by the switches with the same number.

1.6 (a) 1 0 1 0 1 0 1 0.   (b) 1 1 1 1 1 1 1 1.
(c) 1 1 1 1 1 1 1 1.

Practical Exercise 1.3

Step	Address	Accumulator	Memory address 180B	Output LED's (binary)
0	1800	X	0F	0 0 0 0 0 0 0 0
1	1803	0F	0F	0 0 0 0 0 0 0 0
2	1805	0F	0F	0 0 0 0 1 1 1 1
3	1807	77	0F	0 0 0 0 1 1 1 1
4	180A	77	77	0 0 0 0 1 1 1 1
5	180B	77	77	0 0 0 0 1 1 1 1

Practical Exercise 1.4

The system memory map is shown in *Figure A2.1*. Each ROM occupies 4 K and the RAM space is 2 k bytes.

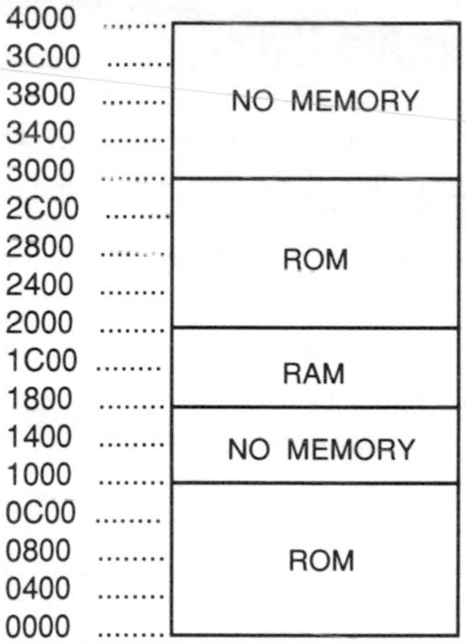

Figure A2.1   Memory map

CHAPTER 2

2.1 All of the address lines show activity:

A_{15}	–	logic 0
A_{14}	–	logic 0
A_{13}	–	logic 0
A_{12}	–	pulse activity
A_{11}	–	pulse activity
A_{10}	–	logic 0
A_9	–	logic 0
A_8	–	logic 0
A_7	–	logic 0
A_6	–	pulse activity
A_5	–	pulse activity
A_4	–	pulse activity
A_3	–	pulse activity
A_2	–	pulse activity
A_1	–	pulse activity
A_0	–	pulse activity

D_7	–	pulse activity
D_6	–	pulse activity
D_5	–	pulse activity
D_4	–	pulse activity
D_3	–	pulse activity
D_2	–	pulse activity
D_1	–	pulse activity
$\underline{D_0}$	–	pulse activity
$\overline{MREQ}$	–	pulse activity
$\overline{IORQ}$	–	logic 1 – no input/output instructions.
$\overline{RD}$	–	pulse activity
$\overline{WR}$	–	logic 1 – no write instructions.

2.2 Since the only instruction in the program is a jump, then this is continually **read** from memory. No writing or input/output is involved, so these control lines are inactive.

2.3 The pulses on $\overline{WR}$ and $\overline{IORQ}$ are at twice the rate of the port 81H lights. This is because new data is written to the output for each state, which requires two bytes of input data for each cycle of the waveform. $\overline{MREQ}$ and $\overline{RD}$ pulse continuously since the program is being **read** from memory.

2.4 These pins pulse at the same rate as the port 81H lights. They must be the connections of the output port on the PIO chip.

2.5

Registers								Memory	
PC	A	F	B	C	D	E	H	L	1B00	1B20
1800	00	00	00	00	00	00	00	00	00	00
1803	00	00	00	00	00	00	1B	00	00	00
1805	00	00	00	00	00	00	1B	00	25	00
1806	00	00	25	00	00	00	1B	00	25	00
1807	25	00	25	00	00	00	1B	00	25	00
180A	25	00	25	00	1B	20	1B	00	25	00
180C	75	00	25	00	1B	20	1B	00	25	00
180D	75	00	26	00	1B	20	1B	00	25	00
180E	75	00	26	00	1B	20	1B	00	25	75
180F	75	00	26	00	1B	20	1B	00	26	75

2.6 The program must be:

Address	Hex code	Mnemonic
1800	OE 17	LD C,17H
1802	06 78	LD B,78H
1804	21 31 18	LD HL,1831H
1807	70	LD (HL),B
1808	79	LD A,C
1809	11 30 18	LD DE,1830
180C	12	LD (DE),

Practical Exercise 2.3

(f) The program increments the binary data on the lights.

2.7 Breakpoints could have been set at addresses 1802, 1803, 1805, 1806 or 1807 hex.

2.8 The program shifts the data on the port one place to the left on each loop.

2.9

PC	A	B	C	D	E	H	L	Port 81H
1800	00	00	00	00	00	00	00	00
1802	00	20	00	00	00	00	00	00
1805	00	20	00	00	00	19	00	00
1806	(00)	20	00	00	00	19	00	00
1807	00	20	00	00	00	19	01	(00)
1809	00	20	00	00	00	19	01	00
180C	00	20	00	10	00	19	01	00
180D	00	20	00	0F	FF	19	01	00
180E	0F	20	00	0F	FF	19	01	00
180F	FF	20	00	0F	FF	19	01	00
			DELAY					
1812	00	20	00	00	00	19	01	00
1813	00	1F	00	00	00	19	01	00
			LOOP					
00	00	00	00	00	00	19	20	F8
			REPEAT					

Practical Exercise 2.4

Table 2.1 in the experiment should look like this:

Step	Address bus	Data bus	Control bus	Cycle type
1	1800	3A	M1 RD MREQ	Instruction fetch
2	1801	0B	RD MREQ	Memory read
3	1802	18	RD MREQ	Memory read
4	180B	0F	RD MREQ	Memory read
5	1803	D3	M1 RD MREQ	Instruction fetch
6	1804	81	RD MREQ	Memory read
7	0F81	0F	WR IORQ	Write output
8	1805	DB	M1 RD MREQ	Instruction fetch
9	1806	80	RD MREQ	Memory read
10	0F80	77	RD IORQ	Read input
11	1807	32	M1 RD MREQ	Instruction fetch
12	1808	0B	RD MREQ	Memory read
13	1809	18	RD MREQ	Memory read
14	180B	77	WR MREQ	Memory write
15	180A	76	M1 RD MREQ	Instruction fetch
16				

2.10 LD A,(180BH) – 4 machine cycles
OUT (81H),A – 3 machine cycles
IN A,(80H) – 3 machine cycles
LD (180BH),A – 4 machine cycles
HALT – 1 machine cycle

2.11 The port address appears twice in each input and output instruction, once on the data bus during the second machine cycle and then on the lower half of the address bus during the third machine cycle in the instruction execution.

2.12 (a) See *Figure A2.2(a)*.
(b) See *Figure A2.2(b)* overleaf.

CHAPTER 3

Practical Exercise 3.1

(b) 57 + 1B = 72 hex

(e) (i) 29 + 55 in decimal is 0 1 0 1 0 1 0 0 in binary. Result is 54 hex or 84 in decimal.
(ii) −10 + 93 in decimal is 0 1 0 1 0 0 1 1 in binary. Result is 53 hex or 83 in decimal.
(iii) −17 + (−34) in decimal is 1 1 0 0 1 1 0 1 in binary. Result is CD in hex or −51 in decimal.

(f) (i) With carry = 0, sum is 1 1 1 1 1 1 1 1. With carry = 1, sum is 0 0 0 0 0 0 0 0.

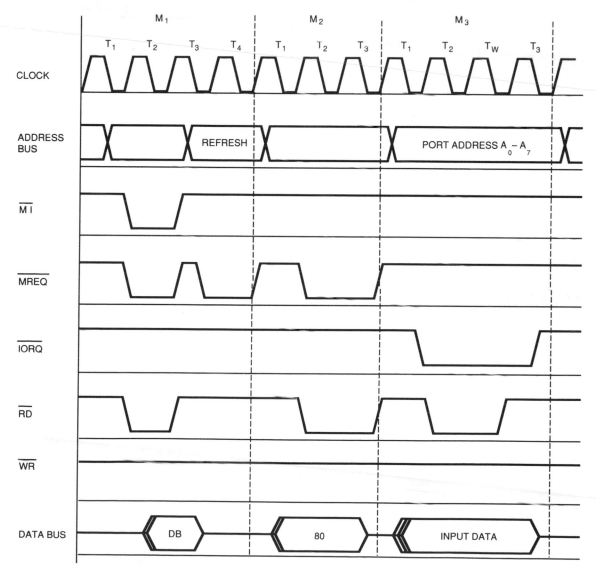

Figure A2.2(a) Answer to Q2.12(a)

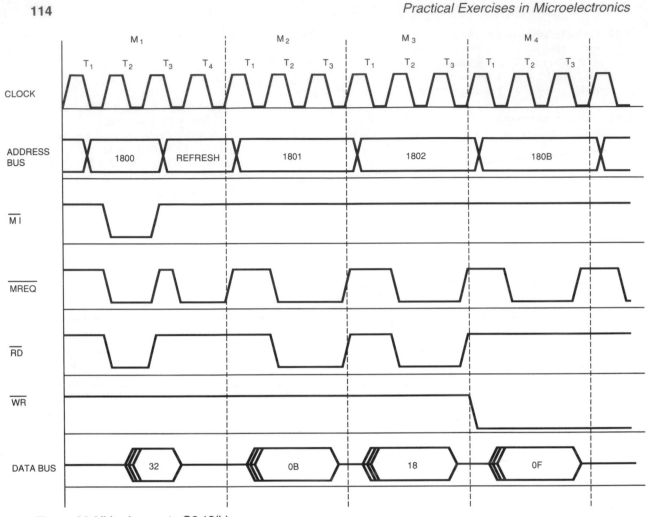

Figure A2.2(b)   Answer to Q2.12(b)

(ii) With carry = 0, sum is 0 0 0 0 0 0 0 0.
      With carry = 1, sum is 0 0 0 0 0 0 0 1.
(iii) With carry = 0, sum is 1 1 1 0 0 0 0 1.
      With carry = 1, sum is 1 1 1 0 0 0 1 0.

3.1 Replace the instruction LD (HL),A with OUT
(81H),A.

3.2 Indirect addressing could not be used. The new
program would become:

> LD A,B
> ADD A,C
> LD D,A
> HALT

(g) (i) 0 1 0 0 1 1 1 0.

(ii) 1 0 1 1 1 1 1 1 − Carry flag is set.
(iii) 1 0 0 0 0 0 0 0 − Carry flag is set.

3.3

Address	Hex code	Mnemonic
		ORG 1800H
1800	DB 80	IN A,(80H)
1802	47	LD B,A
1803	3A 02 00	LD A,(0002H)
1806	90	SUB B
1807	32 00 1A	LD (1A00H),A
180A	CE 00	ADC A,00
180C	32 01 1A	LD (1A01H),A
180F	76	HALT

The lowest input number is C4 hex.

Practical Exercise 3.2

(a)

Address	Hex code		Mnemonic
			ORG 1800H
1800	3E 05	START:	LD A,05
1802	1E 03		LD E,03
1804	CD 00 19		CALL 1900H
1807	76		HALT

3.4 The program should work with all 8-bit numbers.

3.5 (A)

Address	Hex code		Mnemonic
			ORG 1800H
1800	3E 10	START:	LD A,16
1802	IE 07		LD E,7
1804	CD 00 19		CALL 1900H
1807	3E 02		LD A,02
1809	5D		LD E,L
180A	CD 00 19		CALL 1900H
180D	3E 6D		LD A,109
180F	5D		LD E,L
1810	CD 00 19		CALL 1900H
1813	76		HALT

The result should be 5F60H or 24416 decimal.

(B)

Address	Hex code		Mnemonic
			ORG 1800H
1800	11 34 12	START:	LD DE,1234
1803	01 23 01		LD BC,0123
1806	CD 20 19		CALL DIVIS
1809	76		HALT

3.6 Yes, the result is all 1's.

3.7 (a) 10H remainder 04H.
(b) A3DH remainder 0AH.
(c) 0000H remainder 00H.
(d) 0000H remainder 5000H.

3.8

Address	Hex code		Mnemonic
			ORG 1800H
1800	11 00 90	START:	LD DE,9000H
1803	01 20 08		LD BC,820H
1806	CD 20 19		CALL DIVIS
1809	54		LD D,H
180A	5D		LD E,L
180B	01 13 00		LD BC,13
180E	CD 20 19		CALL DIVIS
1811	76		HALT

Practical Exercise 3.3

(a)

Address	Hex code		Mnemonic
			ORG 1800H
1800	AF		XOR A
1801	06 04		LD B,04
1803	11 00 1A		LD DE,1A00H
1806	21 00 1A		LD HL,1A10H
1809	1A	LOOP:	LD A,(DE)
180A	8E		ADC A,(HL)
180B	12		LD (DE),A
180C	13		INC DE
180D	23		INC HL
180E	05		DEC B
180F	C2 09 18		JP NZ,LOOP
1812	76		HALT

(b) 3DAAB03F hex.

3.9 (a) BBBBBBBB hex.
(b) 5000003 hex.
(c) 3D280 hex.

3.10 The carry flag is SET.

3.11 256 pairs – it depends upon the number capable of being stored in the byte counter register.

3.12 (a) 24196BDE hex.
(b) FFFFFFFF hex – Carry set.
(c) C2EFE hex.

3.13

Address	Hex code		Mnemonic
			ORG 1800H
1800	0E 05		LD C,05H
1802	21 10 1A		LD HL,1A10H
1805	11 00 1A	REPT:	LD DE,1A00H
1808	06 04		LD B,04H
180A	AF		XOR A
180B	1A	LOOP:	LD A,(DE)
180C	8F		ADC A,(HL)
180D	12		LD (DE),A
180E	13		INC DE
180F	23		INC HL
1810	05		DEC B
1811	C2 0B 18		JP NZ,LOOP
1814	7D		LD A,L
1815	C6 0C		ADD A,0CH
1817	6F		LD L,A
1818	0D		DEC C
1819	C2 05 18		JP NZ,REPT
181C	76		HALT

Practical Exercise 3.4

(A) Operation of the DAA instruction

(d)

n	m	Hex	Decimal sum	Correction sum	
44	72	B6	(1)16	60	
	80	19	99	99	00
	88	08	90	96	06
	67	35	9C	(1)02	66

(e)

n	m	Hex	Decimal difference	Correction difference	
	76	28	4E	48	FA
	82	41	41	41	00
	11	22	EF	89	9A
	33	42	F1	91	A0

3.14 The carry flag is *set* after:
44 + 72
67 + 35
11 − 22
and 33 − 42

Practical Exercise 3.4

(B) Multi-byte decimal addition

(b) The result is 380631.

(d) The result is 522766.

(e) (i) 008581.
   (ii) 885045.
   (iii) 3091.

3.15 The program works equally well with subtraction.

CHAPTER 4

Practical Exercise 4.1

(A) Simple motor control

(g) Nothing happens without running the initialisation program.

4.1 OR 80H could be made to force bit 7 to logic 1. OR 40H would make it reverse if used in program 1.

(B) Creating a time delay

The program takes 0.88 seconds to execute.

(a)

Address	Hex code		Mnemonic
			ORG 1800H
1800	3E FF	INIT:	LD A,0FFH
1802	D3 82		OUT (82H),A
1804	3E 3F		LD A,3FH
1806	D3 82		OUT (82H),A
1808	3E 80	FWD:	LD A,80H
180A	D3 80		OUT (80H),A
180C	01 00 00		LD BC,0000
180F	0B	DEL:	DEC BC
1810	78		LD A,B
1811	B1		OR C
1812	C2 0F 18		JP NZ,DEL
1815	3E 00	STOP:	LD A,00H
1817	D3 80		OUT (80H),A
1819	76		HALT

(c) The instruction XOR 0C0H inverts bits 6 and 7.

(d)

Address	Hex code		Mnemonic
			ORG 1800H
1800	3E FF	INIT:	LD A,0FFH
1802	D3 82		OUT (82H),A
1804	3E 3F		LD A,3FH
1806	D3 82		OUT (82H),A
1808	3E 80		LD A,80H
180A	57	TURN:	LD D,A
180B	D3 80		OUT (80H),A
180D	01 00 00		LD BC,0000H
1810	0B	DEL:	DEC BC
1811	78		LD A,B
1812	B1		OR C
1813	C2 10 18		JP NZ,DEL
1816	7A		LD A,D
1817	EE C0		XOR C0H
1819	C3 0A 18		JP TURN

4.2

Address	Hex code		Mnemonic
			ORG 1800H
1800	3E FF	INIT:	LD A,0FFH
1802	D3 82		OUT (82H),A
1804	3E 3F		LD A,3FH
1806	D3 82		OUT (82H),A
1808	3E 80		LD A,80H
180A	57	TURN:	LD D,A
180B	D3 80		OUT (80H),A
180D	1E 0A		LD E,0AH
180F	01 00 00	DELAY:	LD BC,0000
1812	0B	DEL:	DEC BC
1813	78		LD A,B
1814	B1		OR C
1815	C2 12 18		JP NZ,DEL
1818	1D		DEC E
1819	C2 0F 18		JP NZ,DELAY
181C	7A		LD A,D
181D	EE C0		XOR 0C0H
181F	C3 0A 18		JP TURN

4.3

Address	Hex code		Mnemonic
			ORG 1800H
1800	3E FF	INIT:	LD A,0FFH
1802	D3 82		OUT (82H),A

Address	Hex code		Mnemonic
1804	3E 3F		LD A,3FH
1806	D3 82		OUT (82H),A
1808	DB 80	LOOP:	IN A,(80H)
180A	0F		RRCA
180B	0F		RRCA
180C	D3 80		OUT (80H),A
180E	C3 08 18		JP LOOP

Practical Exercise 4.2

(A) Testing for certain input conditions

(b) Program 1

Address	Hex code		Mnemonic
			ORG 1800H
1800	3E FF	INIT:	LD A,0FFH
1802	D3 82		OUT (82H),A
1804	3E 3F		LD A,3FH
1806	D3 82		OUT (82H),A
1808	DB 80	START:	IN A,(80H)
180A	E6 04		AND 04H
180C	CA 16 18		JP Z,STOP
180F	3E 80	FWD:	LD A,80H
1811	D3 80		OUT (80H),A
1813	C3 08 18		JP START
1816	3E 00	STOP:	LD A,00H
1818	D3 80		OUT (80H),A
181A	C3 08 18		JP START

(d) Change the instruction at address 180A hex to AND 08H.

(e) Change byte 180C hex from CA to C2 (JP NZ,STOP).

4.4 At address 180A hex, use the instruction AND 0CH.

4.5 Use the instruction AND 0CH in address 180A hex
followed by CP 0CH
and JP NZ,STOP

Practical Exercise 4.2

(B) Waiting for an input to change

(a)

Address	Hex code		Mnemonic
			ORG 1800H
1800	3E FF	INIT:	LD A,0FFH
1802	D3 82		OUT (82H),A
1804	3E 3F		LD A,3FH

Address	Hex code		Mnemonic
1806	D3 82		OUT (82H),A
1808	0E 80		LD C,80H
180A	DB 80	GETDAT:	IN A,(80H)
180C	E6 0F		AND 0FH
180E	47		LD B,A
180F	DB 80	LOOP:	IN A,(80H)
1811	E6 0F		AND 0FH
1813	B8		CP B
1814	47		LD B,A
1815	C2 1B 18		JP NZ,MOTA
1818	C3 0F 18		JP LOOP
181B	79	MOTA:	LD A,C
181C	EE 80		XOR 80H
181E	4F		LD C,A
181F	D3 80		OUT (80H),A
1821	C3 0F 18		JP LOOP

4.6 No. The program still works in the same way since the XOR 80H instruction changes bit 7 of the register whether it is a 1 or 0 initially.

4.7 No. These instructions can also be removed and the program will still work. The first time the program executes the loop it places a value in the B Register.

4.8 It must always be updated with the last input received each time the loop is executed. Since the instruction is from the load group it does not affect the flags and can, therefore, be safely placed between the compare instruction and its associated conditional jump.

4.9

Address	Hex code		Mnemonic
			ORG 1800H
1800	DB 80	START:	IN A,(80H)
1802	E6 78		AND 78H
1804	FE 50		CP 50H
1806	CA 10 18		JP Z,YES
1809	3E 0F		LD A,0FH
180B	D3 81		OUT (81H),A
180D	C3 00 18		JP START
1810	3E F0	YES:	LD A,0F0H
1812	D3 81		OUT (81H),A
1814	C3 00 18		JP START

4.10

Address	Hex code		Mnemonic
			ORG 1800H
1800	3E 00	START:	LD A,00H
1802	D3 81		OUT (81H),A
1804	4F		LD C,A
1805	DB 80		IN A,(80H)
1807	47		LD B,A
1808	DB 80	LOOP:	IN A,(80H)
180A	B8		CP B
180B	47		LD B,A
180C	CA 08 18		JP Z,LOOP
180F	0C		INC C
1810	79		LD A,C
1811	D3 81		OUT (81H),A
1813	C3 08 18		JP LOOP

Procedure

(a)

Address	Hex code		Mnemonic
			ORG 1800H
1800	3E FF	MINI:	LD A,0FFH
1802	D3 82		OUT (82H),A
1804	3E 3F		LD A,3FH
1806	D3 82		OUT (82H),A
1808	C9		RET

(b)

Address	Hex code		Mnemonic
			ORG 1810H
1810	3E 80	FWD:	LD A,80H
1812	D3 80		OUT (80H),A
1814	C9		RET

(c)

Address	Hex code		Mnemonic
			ORG 1820H
1820	3E 40	REV:	LD A,40H
1822	D3 80		OUT (80H),A
1824	C9		RET

(d)

Address	Hex code		Mnemonic
			ORG 1830H
1830	3E 00	STOP:	LD A,00
1832	D3 80		OUT (80H),A
1834	C9		RET

(e)

Address	Hex code		Mnemonic
			ORG 1840H
1840	01 00 00	DELAY:	LD BC,0000
1843	0B	DEL:	DEC BC
1844	78		LD A,B
1845	B1		OR C
1846	C2 43 18		JP NZ,DEL
1849	C9		RET

(f)

Address	Hex code		Mnemonic
			ORG 1850H
1850	06 03	PROP:	LD B,03
1852	DB 80	LOOP1:	IN A,(80H)
1854	E6 10		AND 10H
1856	C2 52 18		JP NZ,LOOP1
1859	DB 80	LOOP0:	IN A,(80H)
185B	E6 10		AND 10H
185D	CA 59 18		JP Z,LOOP0
1860	05		DEC B
1861	C2 52 18		JP NZ,LOOP1
1864	C9		RET

Program 2

(g)

Address	Hex code		Mnemonic
			ORG 1870H
1870	CD 00 18	MAIN:	CALL 1800H
1873	CD 10 18	LOOP:	CALL 1810H
1876	CD 50 18		CALL 1850H
1879	CD 30 18		CALL 1830H
187C	CD 40 18		CALL 1840H
187F	CD 20 18		CALL 1820H
1882	CD 50 18		CALL 1850H
1885	CD 30 18		CALL 1830H
1888	CD 40 18		CALL 1840H
188B	C3 73 18		JP 1873H

4.11 Replace the simple subroutine **call** for 1 revolution with the following program lines:

```
        LD D,05      ; USE D AS COUNTER
REP:    CALL 1850H   ; COUNT 1 REV
        DEC D
        JP NZ, REP
```

An alternative method would be to replace the number 03 in the revolution counting subroutine with the number 0F hex. However, this would then limit the maximum number of revolutions that could be counted.

4.12

Address	Hex code		Mnemonic
			ORG 1900H
1900	CD 00 18	MAIN:	CALL MINI
1903	DB 80	LOOP:	IN A,(80H)
1905	E6 0F		AND 0FH
1907	CA 03 19		JP Z,LOOP
190A	57		LD D,A
190B	CD 10 18		CALL FWD
190E	CD 50 18	REP:	CALL PROP
1911	15		DEC D
1912	C2 0E 19		JP NZ,REP
1915	CD 30 18		CALL STOP
1918	CD 40 18		CALL DELAY
191B	DB 80	LOOP1:	IN A,(80H)
191D	E6 0F		AND 0FH
191F	CA 1B 19		JP Z,LOOP1
1922	57		LD D,A
1923	CD 20 18		CALL REV
1926	CD 50 18	REP1:	CALL PROP
1929	15		DEC D
192A	C2 26 19		JP NZ,REP1
192D	CD 30 18		CALL STOP
1930	CD 40 18		CALL DELAY
1933	C3 03 19		JP LOOP

Practical Exercise 4.4

(A) Program 1

Address	Hex code		Mnemonic
			ORG 1800H
1800	CD 9C 20	THERM:	CALL MHINI
1803	CD 01 20	LOOP:	CALL ANALOG
1806	FE 80		CP 80H
1808	DA 14 18		JP C,HON
180B	F5	HOFF:	PUSH AF

Address	Hex code		Mnemonic
180C	3E 00		LD A,00
180E	D3 80		OUT (80H),A
1810	F1		POP AF
1811	C3 1A 18		JP DIS
1814	F5	HON:	PUSH AF
1815	3E 20		LD A,20H
1817	D3 80		OUT (80H),A
1819	F1		POP AF
181A	CD 20 24	DIS:	CALL DISPA
181D	C3 03 18		JP LOOP

Address	Hex code		Mnemonic
181C	F5	HOFF:	PUSH AF
181D	3E 00		LD A,00
181F	D3 80		OUT (80H),A
1821	F1		POP AF
1822	C3 2B 18		JP DIS
1825	F5	HON:	PUSH AF
1826	3E 20		LD A,20H
1828	D3 80		OUT (80H),A
182A	F1		POP AF
182B	CD 20 24	DIS:	CALL DISPA
182E	C3 03 18		JP LOOP

4.13 The highest and lowest temperatures recorded may vary between applications boards but should not be more than 82H for the highest value and 7DH for the lowest.

4.14 This also varies between applications boards. Typically the highest threshold is about C8H. At this point the heater cannot get any hotter because it is being cooled in the air. The ambient temperature will also affect the result.

The lowest threshold limit should be the Hex equivalent of the ambient temperature since the heater cannot produce any cooling effect.

Practical Exercise 4.4 (B)

(a) A suitable machine code program is given below:

Address	Hex code		Mnemonic
			ORG 1800H
1800	CD 9C 20	THERM2:	CALL MHINI
1803	CD 01 20	LOOP:	CALL ANALOG
1806	FE 80		CP 80H
1808	DA 25 18		JP C,HON
180B	FE 90		CP 90H
180D	DA 1C 18		JP C,HOFF
1810	F5	OVER:	PUSH AF
1811	3E 00		LD A,00
1813	D3 80		OUT (80H),A
1815	F1		POP AF
1816	CD 5E 24		CALL ALARM
1819	C3 03 18		JP LOOP

4.15

Address	Hex code		Mnemonic
			ORG 1800H
1800	CD 9C 20	THERM3:	CALL MHINI
1803	CD 01 20	LOOP:	CALL ANALOG
1806	FE 70		CP 70H
1808	DA 36 18		JP C,UNDER
180B	FE 80		JP 80H
180D	DA 2A 18		JP C,HON
1810	FE 90		CP 90H
1812	DA 21 18		JP C,HOFF
1815	F5	OVER:	PUSH AF
1816	3E 00		LD A,00
1818	D3 80		OUT (80H),A
181A	F1		POP AF
181B	CD 5E 24	ALAM:	CALL ALARM
181E	C3 03 18		JP LOOP
1821	F5	HOFF:	PUSH AF
1822	3E 00		LD A,00
1824	D3 80		OUT (80H),A
1826	F1		POP AF
1827	C3 30 18		JP DIS
182A	F5	HON:	PUSH AF
182B	3E 20		LD A,20H
182D	D3 80		OUT (80H),A
182F	F1		POP AF
1830	CD 20 24	DIS:	CALL DISPA
1833	C3 03 18		JP LOOP
1836	F5	UNDER:	PUSH AF
1837	3E 20		LD A,20H
1839	D3 80		OUT (80H),A
183B	F1		POP AF
183C	C3 1B 18		JP ALAM

CHAPTER 5

5.1 The time for each cycle of the waveform changes.

5.2 Change the number of values to be output by replacing LD B,20H with LD B,10H.

5.3 Read the Table in reverse by changing LD HL,1900H to LD HL,191FH; and INC HL to DEC HL.

5.4 The state table, entered into addresses starting at 1900H would be:

1900	FF
1901	FF
1902	00
1903	00
1904	00

The minimum delay period must be 0.1 seconds.

5.5 The sine wave data table with 36 values is shown below:

1900	80	0 degrees
1901	96	
1902	AB	
1903	C0	
1904	D2	
1905	E1	
1906	EE	
1907	F7	
1908	FD	
1909	FF	90 degrees
190A	FD	
190B	F7	
190C	EE	
190D	E1	
190E	D2	
190F	C0	
1910	AB	
1911	96	
1912	80	180 degrees
1913	6A	
1914	55	
1915	41	
1916	2E	
1917	1F	
1918	12	
1919	09	
191A	03	
191B	01	270 degrees
191C	03	
191D	09	
191E	12	
191F	1F	
1920	2E	
1921	41	
1922	55	
1923	6A	

Practical Exercise 5.2

(A) Traffic light sequence

Time	7	6	5	4	3	2	1	0	Data code	Time code
1	1	0	0	0	1	1	0	0	8C	02
1	1	0	0	0	0	0	1	0	82	02
4	1	0	0	1	0	0	1	0	92	08
0.5	1	0	0	0	0	0	1	0	82	01
0.5	1	0	0	1	0	0	1	0	92	01
0.5	1	0	0	0	0	0	1	0	82	01
0.5	1	0	0	1	0	0	1	0	92	01
0.5	1	0	0	0	0	0	1	0	82	01
0.5	1	0	0	1	0	0	1	0	92	01
1	1	0	0	0	0	1	0	0	84	02
1	1	1	0	0	1	0	0	0	C8	02
1	0	0	1	0	1	0	0	0	28	02
4	0	0	1	0	1	0	0	1	29	08
0.5	0	0	1	0	1	0	0	0	28	01
0.5	0	0	1	0	1	0	0	1	29	01
0.5	0	0	1	0	1	0	0	0	28	01
0.5	0	0	1	0	1	0	0	1	29	01
0.5	0	0	1	0	1	0	0	0	28	01
0.5	0	0	1	0	1	0	0	1	29	01
1	0	1	0	0	1	0	0	0	48	02

The data and time codes occupy alternate memory addresses starting at 1A00H.

Program 2

Address	Hex code		Mnemonic
			ORG 1800H
1800	06 14		LD B,14H
1802	21 00 1A		LD HL,1A00H
1805	7E	REP:	LD A,(HL)
1806	23		INC HL
1807	D3 81		OUT (81H),A
1809	4E		LD C,(HL)
180A	23		INC HL
180B	CD A0 18		CALL VARDEL
180E	05		DEC B

Address	Hex code		Mnemonic
180F	C2 05 18		JP NZ,REP
1812	C3 00 18		JP START
			ORG 18A0H
18A0	11 50 91	VARDEL:	LD DE,9150H
18A3	1B	VLOOP:	DEC DE
18A4	7A		LD A,D
18A5	B3		OR E
18A6	C2 A3 18		JP NZ,VLOOP
18A9	0D		DEC C
18AA	C2 A0 18		JP NZ,VARDEL
18AD	C9		RET

(B) Pelican crossing sequence

(a)

Time	7	6	5	4	3	2	1	0	Data code	Time code
5	1	0	0	0	0	0	1	0	82	0A
2	1	0	0	0	0	1	0	0	84	04
1	1	0	0	0	1	0	0	0	88	02
6	0	0	1	0	1	0	0	0	28	0C
0.5	0	0	1	0	0	1	0	0	24	01
0.5	0	0	0	0	0	0	0	0	00	01
0.5	0	0	1	0	0	1	0	0	24	01
0.5	0	0	0	0	0	0	0	0	00	01
0.5	0	0	1	0	0	1	0	0	24	01
0.5	0	0	0	0	0	0	0	0	00	01
0.5	0	0	1	0	0	1	0	0	24	01
0.5	0	0	0	0	0	0	0	0	00	01
0.5	1	0	0	0	0	0	1	0	82	01

(b)

Address	Hex code		Mnemonic
			ORG 1800H
1800	DB 00	WAIT0:	IN A,(00)
1802	E6 40		AND 40H
1804	C2 00 18		JP NZ,WAIT0
1807	DB 00	WAIT1:	IN A,(00)
1809	E6 40		AND 40H
180B	CA 07 18		JP Z,WAIT1
180E	06 0D		LD B,0DH
1810	21 00 1A		LD HL,1A00H
1813	7E	REP:	LD A,(HL)
1814	23		INC HL

Address	Hex code	Mnemonic
1815	D3 81	OUT (81H),A
1817	4E	LD C,(HL)
1818	23	INC HL
1819	CD A0 18	CALL VARDEL
181C	05	DEC B
181D	C2 13 18	JP NZ,REP
1820	C3 00 18	JP WAIT0

5.6 The speaker should bleep when the 'Pedestrian Green' light is showing. This would entail breaking the sequence into two parts, one for the first four steps and another for the remaining steps. The time on Step 4 would have to be changed to 0.5 seconds and it would then be replaced by the bleep subroutine.

(C) Sound sequence

Address	Data	
1A00	2C 01	C
1A02	01	
1A03	64 00	
1A05	01	
1A06	2C 01	
1A08	01	
1A09	64 00	
1A0B	03	
1A0C	2C 01	0
1A0E	01	
1A0F	2C 01	
1A11	01	
1A12	2C 01	
1A14	03	
1A15	2C 01	D
1A17	01	
1A18	64 00	
1A1A	01	
1A1B	64 00	
1A1D	03	
1A1E	64 00	E
1A20	05	

(b)

Address	Hex code		Mnemonic
			ORG 1800H
1800	06 0B	START:	LD B,0BH
1802	21 00 1A		LD HL,1A00H
1805	5E	REP:	LD E,(HL)

Address	Hex code		Mnemonic
1806	23		INC HL
1807	56		LD D,(HL)
1808	23		INC HL
1809	C5		PUSH BC
180A	CD 7C 24		CALL TONE1
180D	C1		POP BC
180E	4E		LD C,(HL)
180F	23		INC HL
1810	CD A0 18		CALL VARDEL
1813	05		DEC B
1814	C2 05 18		JP NZ,REP
1817	C3 00 18		JP START

5.7 The length of the message is limited by the number of states stored in the B register. Thus the maximum is 256. It could be extended if a register pair were used as the counter to 65536.

5.8 About 8.

Practical Exercise 5.3

(a)

Step	Action 7 6 5 4 3 2 1 0	Terminating condition 7 6 5 4 3 2 1 0
1	0 0 0 0 0 0 0 0	0 0 1 0 0 1 0 0
2	0 0 0 1 0 0 0 1	0 0 1 0 0 0 1 0
3	0 0 1 0 0 0 0 1	0 0 1 0 0 0 1 1
4	1 0 0 0 0 0 1 1	0 0 1 1 0 0 1 0
5	0 0 0 0 0 1 0 1	0 0 1 0 0 1 0 0
6	0 0 0 0 1 0 0 1	0 0 1 0 0 0 1 0
7	1 0 0 0 0 0 1 1	0 0 1 0 1 0 1 0
8	0 0 0 0 0 1 0 1	0 0 1 0 0 1 0 0
9	0 1 0 0 0 1 1 1	0 0 1 0 1 1 0 0
10	0 0 0 0 0 0 0 0	0 0 0 0 0 1 0 0

This gives the state table.

1A00	00	24
1A02	11	22
1A04	21	23
1A06	83	32
1A08	05	24
1A0A	09	22
1A0C	83	2A
1A0E	05	24
1A10	47	2C
1A12	00	04

(b)

Address	Hex code		Mnemonic
			ORG 1800H
1800	06 0A	START:	LD B,0AH
1802	21 00 1A		LD HL,1A00H
1805	7E	REP:	LD A,(HL)
1806	D3 81		OUT (81H),A
1808	23		INC HL
1809	4E		LD C,(HL)
180A	23		INC HL
180B	DB 80	LOOP:	IN A,(80H)
180D	B9		CP C
180E	C2 0B 18		JP NZ,LOOP
1811	05		DEC B
1812	C2 05 18		JP NZ,REP
1815	76		HALT

5.9 By including the instruction AND 3FH after the input instruction IN A,(80H).

5.10 (a) Automatic test equipment.
     (b) Computer controlled machine tools.

5.11 The state table would be:

	Output	Condition
1A00	00	81
1A02	01	36
1A04	02	F2
1A06	04	9A
1A08	08	C1
1A0A	10	D5
1A0C	20	0E
1A0E	40	74
1A10	80	00

CHAPTER 6

Practical Exercise 6.1

(a)

Address	Hex code		Mnemonic
			ORG 1900H
1900	1E FE	SCANKY:	LD E,0FEH
1902	0E 00		LD C,0
1904	2E 00		LD L,0
1906	26 06		LD H,6

Address	Hex code		Mnemonic
1908	7B	KCOL:	LD A,E
1909	D3 02		OUT (02),A
190B	00		NOP
190C	06 06		LD B,6
190E	DB 00		IN A,(00)
1910	57		LD D,A
1911	CB 1A	KROW:	RR D
1913	DA 1C 19		JP C,NOKEY
1916	2E 01		LD L,01
1918	79		LD A,C
1919	32 00 1A		LD (1A00H),A
191C	0C	NOKEY:	INC C
191D	05		DEC B
191E	C2 11 19		JP NZ,KROW
1921	CB 03		RLC E
1923	25		DEC H
1924	C2 08 19		JP NZ,KCOL
1927	C9		RET

(b)

Address	Hex code		Mnemonic
			ORG 1800H
1800	CD 00 19	MAIN:	CALL SCANKY
1803	7D		LD A,L
1804	FE 01		CP 01
1806	C2 00 18		JP NZ,MAIN
1809	76		HALT

6.1

Key	Key code
9	0E
A	8
STEP	10
TAPE WR	17
MOVE	23

6.2 The missing codes are 04, 05 0A and 0B.

6.3

Address	Hex code		Mnemonic
			ORG 1800H
1800	CD 00 19	BLEEP:	CALL SCANKY
1803	7D		LD A,L
1804	FE 01		CP 01H
1806	C2 00 18		JP NZ,BLEEP
1809	11 64 00		LD DE,0064H
180C	CD 7C 24		CALL TONE1
180F	C3 00 18		JP BLEEP

6.4

Address	Hex code		Mnemonic
			ORG 1800H
1800	CD 00 19	STEP:	CALL SCANKY
1803	7D		LD A,L
1804	FE 01		CP 01H
1806	C2 00 18		JP NZ,STEP
1809	3A 00 1A		LD A,(1A00H)

(h)

Number of pass through the program loop	PC5 (12)	PC4 (13)	PC3 (17)	PC2 (16)	PC1 (15)	PC0 (14)
1	1	1	1	1	1	0
2	1	1	1	1	0	1
3	1	1	1	0	1	1
4	1	1	0	1	1	1
5	1	0	1	1	1	1
6	0	1	1	1	1	1
7	1	1	1	1	1	0
8	1	1	1	1	0	1

Address	Hex code	Mnemonic
180C	FE 10	CP 10H
180E	C2 00 18	JP NZ,STEP
1811	11 64 00	LD DE,0064H
1814	CD 7C 24	CALL TONE1
1817	C3 00 18	JP STEP

Practical Exercise 6.2

(a)

Display	Segments	d	p	c	b	a	f	g	e	Hex code
1	b,c	0	0	1	1	0	0	0	0	30
2	a,b,d,e,g	1	0	0	1	1	0	1	1	9B
3	a,b,c,d,g	1	0	1	1	1	0	1	0	BA
4	b,c,f,g	0	0	1	1	0	1	1	0	36
5	a,c,d,f,g	1	0	1	0	1	1	1	0	AE
6	a,c,d,e,f,g	1	0	1	0	1	1	1	1	AF

(b)

Address	Hex code		Mnemonic
			ORG 1800H
1800	26 06	MPLEX:	LD H,6
1802	DD 21 00 1A		LD IX,1A00H
1806	1E 01		LD E,01
1808	DD 7E 00	DIG:	LD A,(IX+0)
180B	D3 01		OUT (01),A
180D	7B		LD A,E
180E	F6 C0		OR 0C0H
1810	D3 02		OUT (02),A
1812	CD 30 18		CALL DELAY
1815	3E C0		LD A,0C0H
1817	D3 02		OUT (02),A
1819	DD 23		INC IX
181B	CB 03		RLC E
181D	25		DEC H
181E	C2 08 18		JP NZ,DIG
1821	C3 00 18		JP MPLEX
			ORG 1830H
1830	01 00 00	DELAY:	LD BC,0000H
1833	0B	DEL:	DEC BC
1834	78		LD A,B
1835	B1		OR C
1836	C2 33 18		JP NZ,DEL
1839	C9		RET

6.5 Typically values between about 0100 hex and 0001 hex.

6.6 If the delay time is too short the segments are not illuminated for sufficient time compared with the length of time which the program takes. This means that the display becomes dimmer.

(f) The codes for the messages are:

U	B5
S	AE
L	85
O	BD
P	1F
Blank	00
S	AE
U	B5
P	1F
L	85
E	8F
H	37

6.7 Change LD E,01 to LD E,20H and change RLC E to RRC E.

(B) Using the monitor display subroutine

(a)

Address	Hex code		Mnemonic
			ORG 1800H
1800	DD 21 00 1A	MAIN:	LD IX,1A00H
1804	CD 24 06		CALL SCAN1
1807	C3 00 18		JP MAIN

6.8

Address	Hex code		Mnemonic
			ORG 1800H
1800	DD 21 00 1A	HELLO:	LD IX,1A00H
1804	06 32		LD B,32H
1806	CD 24 06	HLFSEC:	CALL SCAN1
1809	05		DEC B
180A	C2 06 18		JP NZ,HLFSEC
180D	DD 21 A5 07		LD IX,07A5H
1811	06 32	BLANK:	LD B,32H
1813	CD 24 06	BLANK1:	CALL SCAN1
1816	05		DEC B
1817	C2 13 18		JP NZ,BLANK1
181A	C3 00 18		JP HELLO

Address	Hex code		Mnemonic
			ORG 1A00H
1A00	BD		DB 0BDH ; 0
1A01	85		DB 85H ; L
1A02	85		DB 85H ; L
1A03	8F		DB 8FH ; E
1A04	37		DB 37H ; H
1A05	00		DB 00H ; Blank

(c)

Address	Hex code		Mnemonic
			ORG 1800H
1800	0E 2A	MOVER:	LD C,2AH
1802	DD 21 2A 1A		LD IX,1A2AH
1806	06 28	MOVE:	LD B,28H
1808	CD 24 06	MOV1:	CALL SCAN1
180B	05		DEC B
180C	C2 08 18		JP NZ,MOV1
180F	DD 2B		DEC IX
1811	0D		DEC C
1812	C2 06 18		JP NZ,MOVE
1815	C3 00 18		JP MOVER

The message is: SHELL OIL IS good but CAStor is bEST.

6.9 The program is the same as that given in (c) above apart from the values loaded into registers C and IX. These should be:

LD C,17H
LD IX,1A17H

The message will be:

1A00	00	00	00	00	00	00	1B
1A08	3F	8F	87	00	03	A3	00
1A10	8F	8F	0F	0F	BD	8D	00
1A18	00	00	00	00	00		

6.10

Address	Hex code		Mnemonic
			ORG 1800H
1800	DD 21 00 1A	FHELP:	LD IX,1A00H
1804	06 32		LD B,32H
1806	CD 24 06	HLFSEC:	CALL SCAN1
1809	DA 11 18		JP C,NOKY
180C	FE 10		CP 10H
180E	CA 2A 18		JP Z,STOP
1811	05	NOKY:	DEC B

Address	Hex code		Mnemonic
1812	C2 06 18		JP NZ,HLFSEC
1815	DD 21 A5 07	BLANK:	LD IX,07A5H
1819	06 32		LD B,32H
181B	CD 24 06	BLANK1:	CALL SCAN1
181E	DA 26 18		JP C,NOKEY
1821	FE 10		CP 10H
1823	CA 2D 18		JP Z,STOP
1826	05	NOKEY:	DEC B
1827	C2 1B 18		JP NZ,BLANK1
182A	C3 00 18	STOP:	HALT

Practical Exercise 6.3

Address	Hex code		Mnemonic
			ORG 1800H
1800	0E n	MAIN:	LD C,n
1802	CD 40 19		CALL BINBCD
1805	76		HALT

(c) (i) 165.   (ii) 146.   (iii) 255.   (iv) 204.

6.11

Address	Hex code		Mnemonic
			ORG 1940H
1940	AF	BIN16:	XOR A
1941	21 00 1A		LD HL,1A00H
1944	77		LD (HL),A
1945	23		INC HL
1946	77		LD (HL),A
1947	23		INC HL
1948	77		LD (HL),A
1949	1E 10		LD E,16
194B	CB 11	LOOP:	RL C
194D	CB 10		RL B
194F	16 03		LD D,03
1951	21 00 1A		LD HL,1A00H
1954	7E	BCDADJ:	LD A,(HL)
1955	8F		ADC A,A
1956	27		DAA
1957	77		LD (HL),A
1958	23		INC HL
1959	15		DEC D
195A	C2 54 19		JP NZ,BCDADJ
195D	1D		DEC E
195E	C2 4B 19		JP NZ,LOOP
1961	C9		RET

6.12 Simply add the decoded versions of their characters to the **table**. It could then be entered with **binary** numbers in A rather than BCD values.

6.13 Convert the HALT at address 197B to RETURN (C9H), then:

Address	Hex code		Mnemonic
			ORG 1800H
		SCAN1:	EQU 0624H
1800	11 00 1A	START:	LD DE,1A00H
1803	21 10 1A		LD HL,1A10H
1806	06 03		LD B,03H
1808	CD 66 19	LOOP:	CALL BCD1
180B	05		DEC B
180C	C2 08 18		JP NZ,LOOP
180F	DD 21 10 1A		LD IX,1A10H
1813	CD 24 06		CALL SCAN1
1816	C3 00 18		JP START

6.14 Assume that the BCD7SG program exists as a subroutine and the **table** has been extended to include all the **hex** characters.

Address	Hex code		Mnemonic
			ORG 1800H
		SCAN1:	EQU 0624H
1800	32 00 1A	START:	LD (1A00H),A
1803	CD 06 19		CALL BCD7SG
1806	DD 21 10 1A		LD IX,1A10H
180A	CD 24 06		CALL SCAN1
180D	CD 00 18		JP START

CHAPTER 7

Practical Exercise 7.1

Program 1

Address	Hex code		Mnemonic
			ORG 1800H
		DISPA:	EQU 2420H
1800	0E 00	ATOD:	LD C,0
1802	79	ATD:	LD A,C
1803	D3 81		OUT (81H),A
1805	00		NOP
1806	00		NOP
1807	00		NOP
1808	DB 80		IN A,(80H)

Address	Hex code		Mnemonic
180A	E6 08		AND 08H
180C	CA 13 18		JP Z,DIS
180F	0C		INC C
1810	C3 02 18		JP ATD
1813	79	DIS:	LD A,C
1814	CD 20 24		CALL DISPA
1817	C3 00 18		JP ATOD

7.1 The range should be from 00 to FF but some applications board controls may only rise to about F3 hex.

7.2 The computer voltage never reaches the analogue input voltage, so bit 3 of port 80H never changes and no display is produced.

7.3 The longest conversion time will be: $256 \times 30 \ \mu s = 7.68$ ms.

7.4 The lowest value should be about 01 hex if the sensor is completely covered. The upper value generated depends entirely upon the brightness of the ambient light. It may go as high as 80 hex in some systems.

7.5 The main drawback is that the conversion time varies with the magnitude of the analogue input. It can also be a relatively long conversion time at almost 8 ms.

Practical Exercise 7.2

Program

Address	Hex code		Mnemonic
			ORG 1800H
		ANALOG:	EQU 2001H
		BINBCD:	EQU 2059H
		HEX7SG:	EQU 0678H
		HEX7:	EQU 0689H
		SCAN1:	EQU 0624H
1800	21 00 00	DVM1:	LD HL,0000H
1803	22 10 1A		LD (1A10H),HL
1806	22 12 1A		LD (1A12H),HL
1809	22 14 1A		LD (1A14H),HL
180C	CD 01 20	DVM2:	CALL ANALOG
180F	32 00 1A		LD (1A00H),A
1812	CD 59 20		CALL BINBCD
1815	3A 01 1A		LD A,(1A01H)
1818	21 12 1A		LD HL,1A12H
181B	CD 78 06		CALL HEX7SG

Address	Hex code		Mnemonic
181E	3A 02 1A		LD A,(1A02H)
1821	21 14 1A		LD HL,1A14H
1824	CD 89 06		CALL HEX7
1827	F6 40	DPOINT:	OR 40H
1829	77		LD (HL),A
182A	DD 21 10 1A		LD IX,1A10H
182E	CD 24 06		CALL SCAN1
1831	C3 0C 18		JP DVM2

(d) (i) The light sensor generates voltages in the range 0.02–2.25 V, depending upon light intensity.

   (ii) The heater voltages are in the range 0.30–1.8 V, depending upon the ambient temperature.

7.6 (a) The range of the DVM is increased to 10 volts. However, its accuracy is reduced to ±80 mV.

   (b) The following instructions should be added between CALL BINBCD and LD A,(1A01H) in the previous program.

```
              LD C,02
   AGAIN:     LD HL,1A01H
              LD B,02
   NEXT:      LD A,(HL)
              RL A
              DAA
              LD (HL),A
              INC HL
              DEC B
              JP NZ,NEXT
              DEC C
              JP NZ,AGAIN
```

Practical Exercise 7.3

Program

(a)

Address	Hex code	Mnemonic
		ORG 1800H
1800	3E FF	LD A,0FFH
1802	DE 83	OUT (83H),A
1804	3E 0F	LD A,0FH
1806	D3 83	OUT (83H),A
1808	3E AA	LD A,0AAH
180A	D3 81	OUT (81H),A
180C	76	HALT

7.7 AF is displayed on the lights.

(c) Change the instruction at address 1804H to LD A,0F0H, then FA is displayed.

7.8 AF is displayed first, then after a delay FA appears.

(d)

Address	Hex code	Mnemonic
		ORG 1800H
1800	3E FF	LD A,0FFH
1802	D3 82	OUT (82H),A
1804	3E 1F	LD A,1FH
1806	D3 82	OUT (82H),A
1808	3E FF	LD A,0FFH
180A	D3 80	OUT (80H),A
180C	DB 80	IN A,(80H)
180E	76	HALT

7.9 When an input instruction reads data from a port that is fully or partially configured as an output, it reads the last data sent to the bit configured as outputs. In this case bits 7, 6 and 5 are in such a condition. Thus the input data in A should be E7H.

7.10

Address	Hex code		Mnemonic
			ORG 1800H
1800	3E FF	INIT:	LD A,0FFH
1802	D3 82		OUT (82H),A
1804	3E DF		LD A,0DFH
1806	D3 82		OUT (82H),A
1808	DB 80	LOOP:	IN A,(80H)
180A	E6 07		AND 07H
180C	FE 07		CP 07H
180E	CA 18 18		JP Z,ON
1811	3E 00	OFF:	LD A,00
1813	D3 80		OUT (80H),A
1815	C3 1C 18		JP READ
1818	3E 20	ON:	LD A,20H
181A	D3 80		OUT (80H),A
181C	DB 80	READ:	IN A,(80H)
181E	D3 81		OUT (81H),A
1820	C3 08 18		JP LOOP

Practical Exercise 7.4

Program

Address	Hex code		Mnemonic
			ORG 1800H
1800	3E FF	INIT:	LD A,0FFH
1802	D3 82		OUT (82H),A
1804	3E 3F		LD A,3FH
1806	D3 82		OUT (82H),A
1808	06 00	COUNT:	LD B,00H
180A	78	LOOP:	LD A,B
180B	FE 00		CP 00
180D	C2 1D 18		JP NZ,OFF
1810	DB 80	GETDAT:	IN A,(80H)
1812	E6 0F		AND 0FH
1814	07		RLCA
1815	07		RLCA
1816	07		RLCA
1817	07		RLCA
1818	57		LD D,A
1819	3E 80	ON:	LD A,80H
181B	D3 80		OUT (80H),A
181D	78	OFF:	LD A,B
181E	BA		CP D
181F	C2 26 18		JP NZ,NEXT
1822	3E 00	STOP:	LD A,00H
1824	D3 80		OUT (80H),A
1826	04	NEXT:	INC B
1827	C3 0A 18		JP LOOP

7.11 It may be slightly different for each motor unit but it should be about 08H.

(d) Make the changes below:

```
GETDAT:   IN A,(80H)
          LD C,A
          AND 0FH
          RLCA
          RLCA
          RLCA
          RLCA
          LD D,A
ON:       LD A,C
          AND 20H
          JP NZ,FWD
REV:      LD A,40H
          JP SEND
FWD:      LD A,80H
SEND:     OUT (80H),A
OFF:      LD A,B   etc.
```

(e) In the original program, make the following changes:

```
GETDAT:   CALL 2001H
          LD D,A
ON:       LD A,80H   etc.
```

7.12 At very low speeds, the power supplied to the motor is only just large enough to overcome the friction and the effects of the permanent magnetic field in the motor.

7.13 Change the original program as follows:

```
GETDAT:   LD A,0FFH
          OUT (82H),A
          OUT (82H),A   ;   ALL BITS INPUTS
          IN A,(80H)
          AND 0E0H
          LD D,A
          LD A,0FFH
          OUT (82H),A
          LD A,3FH
          OUT (82H),A
ON:       LD A,80H   etc.
```

CHAPTER 8

Practical Exercise 8.1

Program

Address	Hex code		Mnemonic
			ORG 1800H
		HEX7SG:	EQU 0678H
		HEX7:	EQU 0689H
		SCAN1:	EQU 0624H
1800	3E FF	PIOINI:	LD A,0FFH
1802	D3 82		OUT (82H),A
1804	3E 3F		LD A,3FH
1806	D3 82		OUT (82H),A
1808	3E 10		LD A,10H
180A	D3 82		OUT (82H),A
180C	3E 97		LD A,97H
180D	D3 82		OUT (82H),A
1810	3E EF		LD A,0EFH
1812	D3 82		OUT (82H),A
;			
1814	3E 1A	CPUINI:	LD A,1AH
1816	ED 47		LD I,A
1818	21 90 18		LD HL,1890H
181B	22 10 1A		LD (1A10H),HL

Address	Hex code		Mnemonic
181E	ED 5E		IM 2
1820	FB		EI
;			
1121	06 0D	CLBUFS:	LD B,0DH
1823	AF		XOR A
1824	21 00 1A		LD HL,1A00H
1827	77	CLB:	LD (HL),A
1828	23		INC HL
1829	05		DEC B
182A	C2 27 18		JP NZ,CLB
;			
182D	3E 80	MSTART:	LD A,80H
182F	D3 80		OUT (80H),A
;			
1831	CD 54 18	REPEAT:	CALL BINCON
;			
1834	21 06 1A	BCD7SG:	LD HL,1A06H
1837	3A 03 1A		LD A,(1A03H)
183A	CD 78 06		CALL HEX7SG
183D	3A 04 1A		LD A,(1A04H)
1840	CD 78 06		CALL HEX7SG
1843	3A 05 1A		LD A,(1A05H)
1846	CD 89 06		CALL HEX7
1849	77		LD (HL),A
;			
184A	DD 21 06 1A	SCAN:	LD IX,1A06H
184E	CD 24 06		CALL SCAN1
1851	C3 31 18		JP REPEAT
;			
1854	F3	BINCON:	DI
1855	06 03		LD B,03
1857	21 03 1A		LD HL,1A03H
185A	36 00	BCL:	LD (HL),0
185C	23		INC HL
185D	05		DEC B
185E	C2 5A 18		JP NZ,BCL
1861	2A 01 1A		LD HL,(1A01H)
1864	E5		PUSH HL
1865	0E 10		LD C,10H
1867	21 01 1A	LOOP:	LD HL,1A01H
186A	CB 16		RL (HL)
186C	23		INC HL
186D	CB 16		RL (HL)
186F	23		INC HL
1870	06 03		LD B,03
1872	7E	BCDADJ:	LD A,(HL)
1873	8F		ADC A,A
1874	27		DAA
1875	77		LD (HL),A
1876	23		INC HL
1877	10 F9		DJNZ BCDADJ
1879	0D		DEC C
187A	20 EB JR		NZ,LOOP

Address	Hex code		Mnemonic
187C	EI		POP HL
187D	22 01 1A		LD (1A01H),A
1880	FB		EI
1881	C9		RET
;			
			ORG 1890H
1890	F5	ISR:	PUSH AF
1891	E5		PUSH HL
1892	21 00 1A		LD HL,1A00H
1895	34		INC (HL)
1896	7E		LD A,(HL)
1897	FE 03		CP 03
1899	C2 A7 18		JP NZ,NOT3
189C	2A 01 1A		LD HL,(1A01H)
189F	23		INC HL
18A0	22 91 1A		LD (1A01H),HL
18A3	AF		XOR A
18A4	32 00 1A		LD (1A00H),A
18A7	EI	NOT3:	POP HL
18A8	FI		POP AF
18A9	FB		EI
18AA	ED 4D		RETI

8.1 Both about the same.

8.2 So that the binary data is not modified while it is being translated to BCD format.

8.3 Replace the JP REPEAT instruction with the following JP TEST where TEST is a routine at address 18A0H.

Address	Hex code		Mnemonic
			ORG 18A0H
18A0	3A 01 1A	TEST:	LD A,(1A01H)
18A3	FE 64		CP 100
18A5	DA 31 18		JP C,REPEAT
18A8	3E 00		LD A,00
18AA	D3 80		OUT (80H),A
18AC	C3 31 18		JP REPEAT

Practical Exercise 8.2

Program 1

Address	Hex code		Mnemonic
			ORG 1800H
		HEX7SG:	EQU 0678H
		HEX7:	EQU 0689H
		SCAN1:	EQU 0624H

Address	Hex code		Mnemonic
1800	3E 19	CPUINI:	LD A,19H
1802	ED 47		LD I,A
1804	21 B0 18		LD HL,18B0H
1807	22 60 19		LD (1960H),HL
180A	ED 5E		IM2
180C	FB		EI
;			
180D	3E 60	CTCINI:	LD A,60H
180F	D3 40		OUT (40H),A
1811	3E B7		LD A,0B7H
1813	D3 40		OUT (40H),A
1815	3E 46		LD A,46H
1817	D3 40		OUT (40H),A
;			
1819	ED 5F	RWAIT:	LD A,R
181B	6F		LD L,A
181C	26 00		LD H,0
181E	29		ADD HL,HL
181F	29		ADD HL,HL
1820	29		ADD HL,HL
1821	3E 3F		LD A,3FH
1823	B5		OR L
1824	6F		LD L,A
;			
1825	E5	WLOOP:	PUSH HL
1826	D1		POP DE
1827	7A		LD A,D
1828	B3		OR E
1829	C2 25 18		JP NZ,WLOOP
;			
182C	3E FF	LEDSON:	LD A,0FFH
182E	D3 81		OUT (81H),A
;			
1830	21 00 00	TIME:	LD HL,0
1833	DB 00	TLOOP:	IN A,(00)
1835	E6 40		AND 40H
1837	C2 33 18		JP NZ,TLOOP
183A	3E B3		LD A,0B3H
183C	D3 40		OUT (40H),A
;			
193E	AF		XOR A
183F	EB	INVHL:	EX DE,HL
1840	21 00 00		LD HL,0
1843	ED 52		SBC HL,DE
1845	CD 57 18	DIS:	CALL HLDISP
1848	DA 45 18		JP C,DIS
184B	FE 16		CP 16H
184D	C2 45 18		JP NZ,DIS
1850	3E 00		LD A,00
1852	D3 81		OUT (81H),A
1854	C3 0D 18		JP CTCINI
1857	22 01 1A	HLDISP:	LD (1A01H),HL

Address	Hex code		Mnemonic
185A	E5		PUSH HL
185B	06 03		LD B,03
185D	21 03 1A		LD HL,1A03H
1860	36 00	BCLEAR:	LD (HL),0
1862	23		INC HL
1863	05		DEC B
1864	C2 60 18		JP NZ,BCLEAR
1867	0E 10		LD C,10H
1869	21 01 1A	COLOOP:	LD HL,1A01H
186C	CB 16		RL (HL)
186E	23		INC HL
186F	CB 16		RL (HL)
1871	23		INC HL
1872	06 03		LD B,03
1874	7E	BADJ:	LD A,(HL)
1875	8F		ADC A,A
1876	27		DAA
1877	77		LD (HL),A
1878	23		INC HL
1879	05		DEC B
187A	C2 74 18		JP NZ,BADJ
187D	0D		DEC C
187E	C2 69 18		JP NZ,COLOOP
;			
1881	21 06 1A	BCD7SE:	LD HL,1A06H
1884	3A 03 1A		LD A,(1A03H)
1887	CD 78 06		CALL HEX7SG
188A	3A 04 1A		LD A,(1A04H)
188D	CD 78 06		CALL HEX7SG
1890	3A 05 1A		LD A,(1A05H)
1893	CD 89 06		CALL HEX7
1896	32 0A 1A		LD (1A0AH),A
;			
1899	21 08 1A		LD HL,1A08H
189C	7E		LD A,(HL)
189D	F6 40		OR 40H
189F	77		LD (HL),A
;			
18A0	DD 21 06 1A	SCAN:	LD IX,1A06H
18A4	CD 24 06		CALL SCANI
18A7	EI		POP HL
18A8	22 01 1A		LD (1A01H),HL
18AB	C9		RET
;			
			ORG 18B0H
18B0	2B	ISR:	DEC HL
18B1	FB		EI
18B2	ED 4D		RETI

8.4 655.36 seconds, since HL holds a 16 bit number decremented every 0.01 seconds.

8.5 The program would have to interrupt every 0.001 seconds. Therefore, the time constant register would have to change to 7.

8.6 Change the first few instructions of the RWAIT section of the program to:

RWAIT:	LD HL,00
DISP:	CALL HLDISP
	JP C,DISP
	CP 16H ; WAIT FOR GO KEY
	JP NZ,DISP
	JP TIME

Practical Exercise 8.3

Program

Address	Hex code		Mnemonic		
			ORG 1800H		
		BINBCD:	EQU 2059H		
		CLRBUF:	EQU 207DH		
		HEX7SG:	EQU 0678H		
		HEX7:	EQU 0689H		
		SCAN1:	EQU 0624H		
;					
1800	CD 7D 20	MAIN:	CALL CLRBUF		
1803	3E 19		LD A,19H		
1805	ED 47		LD I,A	;	INT REG LOAD
1807	21 55 18		LD HL,ISR		
180A	22 22 19		LD (1922),HL	;	START ADD.TABLE
180D	ED 5E		IM2	;	ENTRY FOR CTC
				;	CHANNEL 1
180F	CD 91 20	PIO:	CALL PIOINI		
1812	3E 80		LD A,80H		
1814	D3 80		OUT (80H),A	;	START MOTOR
1816	3E 21	INCNT:	LD A,33		
1818	32 30 19		LD (1930H),A	;	1930 COUNTS INT
181B	3E 20	CTCINI:	LD A,20H		
181D	D3 40		OUT (40H),A	;	INTERRUPT VECT.
181F	3E 57		LD A,57H		
1821	D3 40		OUT (40H),A		
1823	3E 00		LD A,0		
1825	D3 40		OUT (40H),A		
1827	3E B7	TIMER:	LD A,0B7H		
1829	D3 41		OUT (41H),A	;	CHAN.1 CONTROL
182B	3E 46		LD A,70		
182D	D3 41		OUT (41H),A		
;					
182F	FB		EI		
;					
1830	3A 00 1A	LP:	LD A,(1A00H)		
1833	F5		PUSH AF		
1834	CD 59 20		CALL BINBCD		
1837	F1		POP AF		
1838	32 00 1A		LD (1A00H),A		

Address	Hex code		Mnemonic	
183B	3A 01 1A		LD A,(1A01H)	
183E	21 02 19		LD HL,1902H	
1841	CD 78 06		CALL HEX7SG	
1844	3A 02 1A		LD A,(1A02H)	
1847	CD 89 06		CALL HEX7	
184A	77		LD (HL),A	
184B	DD 21 00 19		LD IX,1900H	
184F	CD 24 06		CALL SCAN1	
1852	C3 30 18		JP LP	
;				
1855	F5	ISR:	PUSH AF	
1856	3A 30 19		LD A,(1930H)	
1859	3D		DEC A	
185A	32 30 19		LD (1930H),A	
185D	CA 64 18		JP Z,TIMEUP	
1860	F1		POP AF	
1861	FB		EI	
1862	ED 4D		RETI	
;				
1864	DB 40	TIMEUP:	IN A,(40H)	; GET COUNT
1866	ED 44		NEG	
1868	32 00 1A		LD (1A00H),A	; LD BINARY NO.
186B	3E 57	CTC:	LD A,57H	
186D	D3 40		OUT (40H),A	
186F	3E 00		LD A,0	
1871	D3 40		OUT (40H),A	
1873	3E B7		LD A,B7H	
1875	D3 41		OUT (41H),A	
1877	3E 46		LD A,70	
1879	D3 41		OUT (41H),A	
187B	3E 21		LD A,21H	
187D	32 30 19		LD (1930H),A	
1880	F1		POP AF	
1881	FB		EI	
1882	ED 4D		RETI	

(c) Typical speed is 200 revs per sec. or 12 000 r.p.m.

8.7 If a longer time is chosen, the number of pulses will increase beyond the limit of an 8-bit register and this will require a more complex program. However, the accuracy of the measurement could be increased.

The longer the time, the less responsive the speed display to changes in the motor speed since it tends to produce an averaged effect.

8.8 (a) The accuracy of the time measurement.
(b) The length of time chosen over which to perform the measurement and the consequent multiplication factor that must be used to calculate the number of revs per second.

8.9 Change the line which initialises the count to:
LD A,66   (NB 42H)
Change the interrupt service routine

TIMEUP:	IN A,(40H)	
	NEG	
	SRLA ;	DIVIDE BY 2
	LD (1A00H),A	

The speed display appears much less responsive to changes in speed when the propeller blade is touched. Strange values result if the motor runs at full speed in this mode.

Index